551.2
P

Prager, Ellen J.
Furious earth.

$24.95

DATE			

Furious Earth

Furious Earth

The Science and Nature of Earthquakes, Volcanoes, and Tsunamis

Ellen J. Prager, Ph.D.

with
Kate Hutton, Ph.D.
Costas Synolakis, Ph.D.
Stanley Williams, Ph.D.

McGraw-Hill

New York • San Francisco • Washington, D.C. • Auckland • Bogotá
Caracas • Lisbon • London • Madrid • Mexico City • Milan
Montreal • New Delhi • San Juan • Singapore
Sydney • Tokyo • Toronto

Library of Congress Cataloging-in-Publication Data
Prager, Ellen J.
    Furious Earth:the science and nature of earthquakes,
    volcanoes, and tsunamis/Ellen J. Prager.
        p. cm.
    Includes index.
    ISBN 0-07-135161-2 (alk. paper)
    1.Earthquakes. 2.Volcanoes. 3.Tsunamis. I.Title
QE534.2.P731999   99-048386

McGraw-Hill

*A Division of The **McGraw·Hill** Companies*

1 2 3 4 5 6 7 8 9 0 FGR/FGR 9 0 9 8 7 6 5 4 3 2 1 9

ISBN 0-07-135161-2

This book was set in Zapf Elliptical and Square 721 by TopDesk Publishers' Group. *Printed and bound by* Quebecor Fairfield.

McGraw-Hill books are available at special quantity discounts to use as premiums and sales promotions, or for use in corporate training programs. For more information, please write to the Director of Special Sales, McGraw-Hill, 11 West 19th Street, New York, NY 10011. Or contact your local bookstore.

 This book is printed on recycled, acid-free paper containing a minimum of 50 percent recycled, de-inked fiber.

This book is dedicated to my parents, Larry and Phyllis, and my sister, Kathy, for their endless support, encouragement, and sage advice.

Contents

Introduction

We live on a planet whose fertile land, oxygen-rich atmosphere, and bountiful sea sustain life. Yet, it is the same Earth whose awesome fury can cause devastating loss of life and economic disaster. Explosive volcanic eruptions hurtle burning clouds of gas and ash toward centers of human habitation. Earthquakes violently tear apart the land, turn buildings to rubble, and trigger devastating fires. And tsunamis, waves of gigantic proportion, crash onto shores and wash away entire coastal communities. Because human populations are expanding rapidly, these powerful forces of nature will increasingly threaten lives and property. We will never control the furious Earth, but through our scientific understanding of its nature, we may be able to prevent tragic and costly losses.

Earthquakes, volcanoes, and tsunamis have long wrought fear and fascination in the human mind and inspired myth, legend, and numerous Hollywood disaster movies. Today, advanced technology is enabling us to drill, measure, monitor, sample, and image the Earth and its motions as never before. Our scientific understanding of the Earth's dynamic nature is growing by leaps and bounds and has vastly improved over the last century. Scientists continue to study earthquakes, volcanoes, and tsunamis to unlock their geologic mysteries and to help prepare local populations for their impacts in the future. Some of our success and, tragically, much of our failure to understand and predict the furious Earth have been dramatically illustrated by several recent and catastrophic events.

In 1991, a series of steam blasts and hundreds of small earthquakes alerted scientists and local leaders that Mount Pinatubo in the Philippines was awake and restless. The volcano was intensely monitored around the clock by an international team of experts, and the surrounding towns and villages were evacuated. When Mount Pinatubo finally erupted, it was one of the most explosive blasts of the century, and so great in magnitude that ash dispersed in the upper atmosphere caused global cooling lasting more than a year. But because of excellent detection, warning, and cooperation among all involved, relatively few lives were lost.

Recent evacuations on the small Caribbean island of Montserrat also saved hundreds of lives when a once-dormant volcano came back to life and spewed thick clouds of ash and dust skyward. Compare these events with the 1902 eruption of Mount Pelee on the small Caribbean island of Martinique, in which 30,000 people were killed and an entire city wiped out. Given the warning signs, scientists are now significantly better at forecasting the probability of a volcanic eruption. However, the shape, size, and style of an eruption remain more difficult to predict. And when the warning signs are either absent or unheeded, disaster is imminent.

In 1985, an eruption and massive mud flow from the glacier-covered volcano Nevado Del Ruiz killed an estimated 23,000 people. Prior to the eruption, swarms of small earthquakes were recorded on nearby seismographs. Colombian geologists had recently even completed a map delineating the dangers of a Nevado Del Ruiz eruption. But this information went unheeded by local emergency planners, and thousands became victim to the volcano's unleashed power. And during the 1980 eruption of Mount St. Helens, a pyroclastic flow or superheated avalanche of ash and gas surprised even the geologists studying the site, one of whom died in the blast.

In 1995, a powerful earthquake with a magnitude of 6.9 struck Kobe, Japan. On the basis of previous examples of earthquake damage, the Japanese had invested millions of dollars to build structures that could withstand a major earth-shaking event. Unfortunately, they neglected to consider the potential effects of an earthquake on the soft sediment and landfill that underlay much of the city. When the quake struck, ground motion caused the underlying sediments to liquefy. When liquefaction occurs, even the best-built structures give way. In the Kobe earthquake, more than 5,000 people were killed, and over $200 billion worth of damage was incurred.

In the United States, attention to earthquake hazards has focused mainly on northern California, but in January 1994 an earthquake, 6.7 in magnitude, rocked Northridge and revealed a series of previously unmapped faults in southern California. Although only sixty fatalities occurred in the Northridge quake, economic losses exceeded $40 billion. Historical records show that powerful tremors have also occurred along the U.S. East Coast and even in the interior portions of the continent. And new evidence suggests that the U.S. Pacific Northwest is a likely candidate for a major earthquake in the not-so-distant future. What causes these quakes? Will there be future earthquakes in areas ill prepared for the Earth's fury? And will we ever be able to predict earthquakes?

In 1998, the world observed another dramatic and unexpected demonstration of nature's wrath. A wall of water more than 10 meters (32 feet) high crashed onto the coast of Papua New Guinea, killing thousands and devastating the shoreline. Geologists and oceanographers were stunned that such a powerful and damaging tsunami could form and hit the coast with little warning. To prevent the kind of destructive tsunami impact that occurred in Papua New Guinea, and has in the past wreaked havoc along the shores of Alaska, Oregon, and Hawaii, a tsunami warning system has been established throughout the Pacific Ocean. The system is designed to identify potential triggering events and provide at least some warning to threatened coastal populations. Unfortunately, in Papua New Guinea the triggering event occurred only 29 kilometers offshore and almost instantaneously created a devastatingly huge wave. Until recently, tsunami specialists have focused on tsunamis that impact shorelines distant from their origin, but now they are also concentrating their efforts on locally derived events. Again, it was a powerful and costly lesson.

While scientists and emergency planners today are better able to identify and monitor potential hazards from earthquakes, volcanoes, and tsunamis, prediction of the timing, intensity, and character of events remains elusive. In this book we examine our basic understanding of earthquakes, volcanoes, and tsunamis, look at how scientists are studying these phenomena using the latest in modern technology, and reveal some of the questions that remain unanswered.

Plate tectonics, once criticized as an imaginative and radical theory, is now accepted as fact and explains much of the Earth's dynamic nature and its powerful forces, including the distribution of volcanoes, earthquakes, and tsunamis. In the first section of the book, we discuss this revolutionary idea and how it was discovered.

Volcanologist Dr. Stanley Williams knows firsthand the incredible power and danger volcanoes harbor. During a field trip to a Colombian volcano, Williams and his colleagues were caught at the crater's edge in an unexpected and momentous eruption. Although he sustained life-threatening injuries and witnessed the violent deaths of friends, Williams continues to dedicate his life to the study of volcanoes. In the chapter on volcanoes, we describe our basic understanding of volcanoes, examine the tools of the modern volcanologist, and show how science is able to provide timely warnings that can prevent loss of lives and property during volcanic eruptions.

In the chapter on earthquakes, seismologist Dr. Kate Hutton helps to explain why and how quakes occur, where they are most likely to happen, and how science is aiding the ability to prepare for and hopefully one day predict earthquakes.

And finally, in the chapter on tsunamis, tsunami specialist Dr. Costas Synolakis helps to explain what triggers these towering waves, how scientists use computers to model them, and where we stand with regard to the ultimate tsunami warning system. This chapter, like the rest, is not meant to cover the topic in scientific detail or comprehensive depth, for each of these phenomena could easily be the subject of its own book. A list of excellent sources of further information on the topic of each chapter is provided at the end of the book.

The Earth is a fascinating, dynamic, and contradictory place: powerful yet fragile, violent yet serene, vast yet personal. And in this paradoxical world, we may be awed and fascinated by earthquakes, volcanoes, and tsunamis, but we must also respect them as harbingers of potential disaster. The focus of the book is not on death and destruction, but on what we have learned through research, often in the aftermath of disaster, and how science is helping prevent future catastrophes. The importance and relevance of science to society often goes

Metric Conversions

Scientists most commonly work with the metric system. This table provides some general conversions you may refer to as you read this book:

Distance and Length
1 millimeter (mm) = 0.039 inch
1 centimeter (cm) = 0.393 inch
1 meter (m) = 3.280 feet
1 kilometer (km) = 0.621 mile

Area and Volume
1 square centimeter (cm^2) = 0.155 square inch
1 square meter (m^2) = 10.76 square feet = 1.196 square yards
1 cubic centimeter (cm^3) = 0.061 cubic inch
1 cubic meter (m^3) = 35.314 cubic feet = 1.307 cubic yards
1 liter (L) = 1.056 liquid quarts = 33.814 fluid ounces
1 milliliter (mL) = 0.061 cubic inch

Mass & Weight
1 milligram (mg) = 0.015 grain
1 gram (g) = 0.035 ounce
1 kilogram (kg) = 2.205 pounds = 35.273 ounces
1 metric ton (t) = 1.102 short (net) tons = 2,204.623 pounds

Power and Efficiency
1 kilowatt = 1.3 horsepower

Temperature
Absolute zero
    = 0° Kelvin = –273.16° Celsius = – 459.67° Fahrenheit
Freezing point of water
    = –273° K = 0° C = 32° F
Boiling point of water
    = –373° K = 100° C = 212° F

To convert Celsius to Fahrenheit, multiply by 1.8 and add 32

Chapter 1

The Underlying and Dynamic Earth

Throughout the Earth's history, volcanoes have spewed fire, earthquakes have shaken and ripped apart the land, and towering tsunamis have crashed onto the shore. Humans have long sought to explain these demonstrations of the Earth's fury and to find a means to predict as well as prevent their destructive impacts. In ancient times, theories to explain the powerful forces of nature took the form of myth and religion. The ancient Greeks believed that volcanic eruptions were the work of Hephaestus, the god of fire, and his underground companions, the giants. The Romans called Hephaestus Vulcan, from which name the word *volcano* is derived. Villagers living near active volcanoes sacrificed beautiful maidens, young boys, and valuables to placate the powerful gods and prevent future displays of their ire. But even as sacrifices were made and gods were revered, earthquakes, volcanic eruptions, and tsunamis continued to occur.

Today, we turn to science to explain the Earth's dynamic forces and continue to search for answers to the questions that have plagued humankind for centuries. Much of what we know now about the furious Earth is connected through one overarching—or, more appropriately, underlying—theme, plate tectonics. In the early 1900s, when plate tectonics, then called continental drift, was first proposed, the scientific community reacted with utter disbelief. Amazingly, it is only in the last thirty years that plate tectonics has been widely accepted, and yet it has revolutionized our understanding of the Earth and dramatically changed science in numerous disciplines, including geology, oceanography, chemistry, and biology.

Plate Tectonics Discovered

The unearthing, so to speak, of plate tectonics really began when people started to pay serious attention to the intriguing jigsaw-puzzle fit of the continents. This geographic enigma was considered by Leonardo da Vinci in the 15th century, Francis Bacon and the Dutch mapmaker Abraham Ortelius in the 16th century, and naturalists George Buffon and Alexander von Humboldt in the 18th century. In 1858, geographer Antonio Snider-Pelligrini produced one of the first maps illustrating how the continents may once have fit together. At the time, it was thought that only an extremely powerful earthquake or flood of biblical proportions could possibly have caused the colossal breakup of the once-fused continents. Then, in the early 1900s, Alfred Wegener, a young German meteorologist, proposed a radical, evidence-based theory he called *continental drift*. Wegener suggested that more than 200 million years ago, the continents had been part of a much larger supercontinent, and that some 100 to 150 million years ago, the giant land mass separated, the continents drifted apart, and large ocean basins formed in between.

Wegener based his theory of continental drift on several powerful lines of geologic evidence. The presence of similar fossils and rocks on now-distant shores strengthened the remarkable fit of the coastlines as an indicator of an ancient supercontinent. Wegener proposed an elegantly simple analogy: *"It is just as if we were to refit the torn pieces of a newspaper by matching their edges, and then check whether the lines of print run smoothly across. If they do, there is nothing left but to conclude that the pieces were in fact joined in this way."* For Wegener, fossils and rocks were the print, and the continents the paper. He also discovered that fossils characteristic of one climate were now found in environments of a very different climate. In the hot, arid valleys of Africa he discovered glacial deposits, and in the cold, polar regions he found fossilized ferns indicative of a once-tropical climate. In Antarctica, Wegener also found coal deposits that suggested a previously warm climate and tropical vegetation. He hypothesized that these strange findings could only be explained through continental drift. The climates of these regions must have changed as the continents slowly drifted across the Earth's surface, moving through space as well as time. He also noted that in the Nordic region (Norway, Sweden, and Finland), following the melting of ancient glaciers and the removal of the ice's weight, the underlying land rose (a process now known as glacial rebound). Wegener surmised that if the land could move vertically, it could also move laterally across the Earth's surface; and that mountains formed of folded rock could be evidence of this horizontal motion.

Unfortunately for Wegener, he introduced his theory at a time when most scientists staunchly believed that the continents and oceans were fixed features on the surface of the Earth. Rollin Chamberlain, of the University of Chicago, unequivocally stated his reaction to Wegener's hypothesis: *"Can geology still call itself a science, when it is possible for such a theory as this to run wild?"* Another well-known and respected geologist referred to the theory as *"utter, damned rot"!* Wegener's evidence was largely qualitative, and there was a fatal flaw in his theory. He could not come up with a reasonable explanation for how the continents moved. What was the driving force? Scientists throughout the world, particularly in the United States, considered his theory unproven and quite unbelievable. Without better, more definitive evidence, the theory of continental drift was ignored for more than half a century. But until his death in 1930 during an expedition across the Greenland ice cap, Wegener doggedly pursued evidence to prove his theory.

The geologic key to unraveling the mysteries of plate tectonics would come from advances in technology that allowed, for the first time, detailed study of the ocean floor. It was a breakthrough that has been described as a combination of shrewd analysis and serendipity, otherwise known as dumb luck (Dickinson). Before World War I, technology to study the sea floor was relatively crude, and understanding of the ocean and its sediments was superficial at best. Depths were measured by soundings using a long line or cable, weighted at the bottom, and sediments were collected by difficult and tedious dredging operations. After World War I, primitive sonar systems were developed to measure depth by bouncing a pulse of sound off the bottom and recording its travel time back to the ship, a technique known as echo sounding. Using echo sounding, detailed bathymetric or depth surveys of the sea floor began, and by the 1950s these surveys revealed the presence of an enormous, globe-encircling mountain chain beneath the sea—the mid-ocean ridges. Although the presence of some sort of topographic high, possibly a plateau, had been discovered in the 1920s during the Meteor Expedition, it was not until the 1950s that scientists recognized the true extent and nature of the ridge system. The mid-ocean ridges rise an average of 4,500 meters above the sea floor and wrap around the globe for more than 60,000 kilometers (*color plate 3*). Although undoubtedly the most prominent topographic feature on the planet, the mid-ocean ridge system, amazingly, was not discovered until the mid-20th century. Following World War II, further advances in marine technology and increased interest in the ocean and sea floor led to two more startling discoveries.

Using an instrument called a magnetometer, originally developed to detect submarines, scientists discovered odd magnetic variations on the sea floor. When molten rock containing magnetic particles or minerals cools, its magnetism becomes aligned with the Earth's magnetic field, much like a compass. A magnetometer measures the direction and angle or declination of a rock's magnetism. Today, the magnetic properties of a rock cooled at the surface indicate an alignment to the north. Its declination depends on the latitude where it cools. In the 1950s, oceanographers working with the U.S. Office of Naval Research found a zebralike pattern of magnetic anomalies or variations on the sea floor (*Figure 1.1*). In 1963, Fred Vine and Drummond Matthews of the Canadian Geologic Survey proposed that this striped magnetic pattern was produced by repeated reversals of the Earth's magnetic field.

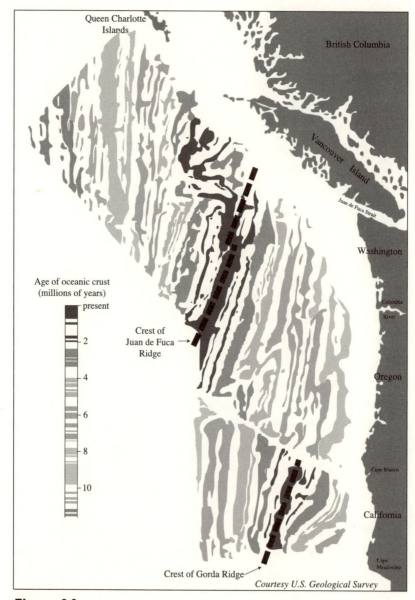

Figure 1.1
Magnetic variations (shaded stripes) and age of the sea floor
in the Pacific Northwest.

Geologists studying rocks on land had previously theorized that throughout the Earth's history its magnetic field had reversed a number of times. Today, compasses point to the north when aligned with the Earth's magnetic field; this is called normal polarity. Following a reversal of the Earth's magnetic field, a compass would point south, thus having a reversed polarity. Scientists today agree that reversals have occurred more than 100 times in the past 75 million years but are still mystified about how and why it happens. Will boaters navigating by compass someday become lost or run aground because the compass needle will suddenly swing south? Or will it start swinging south gradually, and if so, what will initiate the shift? Vine and Matthews suggested that the adjacent stripes on the ocean floor, positive and negative magnetic anomalies (variations), reflected formation during times of normal and reversed polarity, respectively. However, what was truly curious about the magnetic stripes was that they ran parallel to a mid-ocean ridge, and their spacing and width were the same on either side of the ridge's crest.

Just a year earlier, in 1962, Harry Hess of Princeton University, a Wegener fan, speculated that ocean crust is created by volcanism at the crests of mid-ocean ridges, spreads away, and is eventually destroyed in deep ocean trenches. Hess and his colleague Robert Dietz called this process sea-floor spreading, and their work lent support to Vine and Matthews's explanation for magnetic striping. The ocean floor could be seen as a sort of magnetic tape recorder. At a ridge axis, newly forming ocean crust cools and takes on the alignment of the current magnetic field. Spreading of the sea floor then slowly moves the crust away on both sides of the ridge axis. Over time, with repeated reversals of the magnetic field and continued sea-floor spreading, a symmetrical, zebralike pattern forms about the ridge axis (*Figure 1.2*).

The confirmation of sea-floor spreading and its relationship to the zebralike magnetic striping was to come during the early days of the Deep Sea Drilling Project (DSDP). Using a specially designed ship (*color plates 1 and 2*) and some 6,100 meters (20,000 feet) of drill pipe, the DSDP cored and then dated samples of the sea floor. It was an amazing feat, likened to drilling a hole in a New York sidewalk with a strand of spaghetti dangled at night, in swirling winds, from atop the Empire State Building. DSDP data confirmed that with

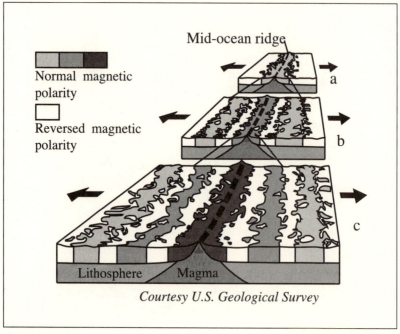

Courtesy U.S. Geological Survey

Figure 1.2
Sea-floor spreading and the creation of magnetic stripes.

increased distance from a ridge axis, the age of the ocean crust increased—the sea floor did indeed form at the axis and spread outward over time (refer to *Figure 1.1*). Two other interesting observations supported Hess's radical idea of sea-floor spreading. Oceanographers and geologists had previously puzzled over why, although the Earth is approximately 4.5 billion years old, the sea floor is much younger, with only a relatively thin layer of sediment and fossils no more than 180 million years old. In 1965, Canadian scientist J. Tuzo Wilson proposed an idea that allowed for the synthesis of continental drift and sea-floor spreading and explained these last two observations.

Wilson, who had been studying earthquakes and faulting in the ocean crust, suggested that the stiff outer rind of the Earth is broken into a number of moving pieces or "plates," and that convergence and destruction of plates at ocean trenches balances spreading at mid-ocean ridges. Fractures across the ridges, which he called transform faults, allowed for motion of the relatively flat plates over the Earth's spherical surface. The distribution of earthquakes and later volcanoes lent credence to the idea that the Earth was divided into numerous platelike sections (*color plate 4*). The destruction of the ocean crust at the trenches could account for the relatively thin sediment cover and young age of the sea floor—the older sea floor and sediments were being recycled back into the Earth.

In the late 1960s, new fossil discoveries were also providing evidence in support of the idea that the continents were shifting and moving about the Earth's surface. In Antarctica, researchers found fossils of a sheep-sized reptile, *Lystrosaurus*, identical to fossils found in both Africa and India dating from 200 million years ago. This indicated that at one time

Antarctica, Africa, and India were connected. Throughout the 1960s and 1970s evidence and support for the theory of plate tectonics mounted, and soon its missing pieces were filled in, including a driving force. The scientific community was given a whole new way of looking at the Earth and the forces that create and mold its surface. What was once only a theory became the accepted dogma, and textbooks everywhere had to be rewritten. We no longer call it continental drift or the plate tectonics theory, but simply plate tectonics. Few other scientific revelations can surpass plate tectonics in helping us understand the present, past, and future shape of the Earth's surface, and its earth-shaking, fiery, and watery displays of power.

Before plate tectonics and studies in modern earth science, people generally thought that the planet was solid throughout its interior and fixed on the surface. But the Earth is now known to be a dynamic sphere whose hot interior moves in slow motion and whose surface shifts and changes with time. The planet's hard surface is relatively thin and composed of a series of interlocking, rigid pieces or plates that move atop a layer of hotter, more fluidlike material—akin to a Milky Way candy bar with a hard, thin layer of brittle chocolate overlying a softer, more fluid caramel layer below. The Earth's internal layering plays a fundamental role in plate tectonics and the planet's furious nature.

Inside the Earth

In 1864, Jules Verne wrote *Journey to the Center of the Earth*, in which his fictional characters embark on a mission to travel into the depths of the Earth and unravel its internal mysteries.

In reality, we have yet to discover a means literally to view the interior of our planet, so we must rely on indirect methods to observe and study its nature. To create a picture of the Earth's internal structure, scientists examine rocks that have been uplifted or ejected from deep within the Earth, drill deep holes into the ocean crust, and measure how seismic waves pass through the Earth's interior. The details of the planet's geologic portrait are still evolving, but for now, the following describes what we think are the Earth's internal composition and physical properties.

The interior of the Earth is composed of several concentric layers (*Figure 1.3*). These layers can be divided in two ways, by

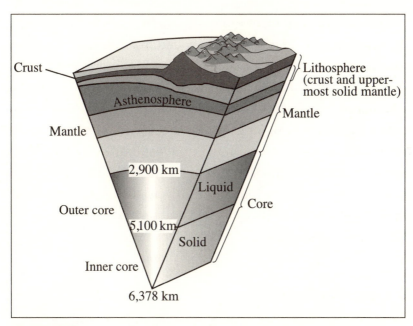

Figure 1.3
Sketch of the Earth's internal concentric layering.
Courtesy of the U.S. Geological Survey.

their chemical composition or their physical properties. Compositionally, the Earth is made up of three main layers, an outer crust (0 to 50 km), an intermediate rocky mantle (50 km to 2,900 km), and an inner metallic core (2,900 to 6,378 km). The outermost layer, the crust, is relatively thin and includes all that we see on land or that lies beneath the sea. Relative to the rest of the Earth, the crust is but a thin sheath covering the planet, much like the outer skin of an apple. The Earth's crust is either oceanic or continental. Compared to oceanic crust, continental crust is less dense, thicker, and composed of lighter minerals. Like icebergs partially hidden beneath the sea's surface, the continents also have an underlying thick root of rock that keeps them sitting high above the underlying layer.

Beneath the Earth's crust lies the rocky mantle, a thick layer of dense, sometimes semimolten, rock, rich in iron and magnesium. A dividing line, called the Mohorovicic discontinuity or simply the Moho, marks the division between the crust and mantle. Based on seismic wave data and analysis of rocks uplifted on land or collected at sea, it is believed that the mantle's composition is similar to that of a mineral called peridiotite (a light- to dark-green silicate [SiO_2] rock with high percentages of magnesium and iron). Below the mantle, at the center of the Earth, some 2,900 kilometers from the surface, is the core. Calculations from measurements of gravity, earthquake data, and the composition of meteorites suggest that the core is composed of a very dense metallic material, probably a mixture of iron and nickel.

Within the Earth, increasing temperature and pressure alters the physical state of its internal layering. Laboratory experiments suggest that the temperature of the outer core hovers around a blazingly hot 5000° C, nearly as hot as the sun's surface (Lamb and Sington). Outside of the core, the Earth's layers insulate the surface from its hot interior, like the lining of a thermos. Much of what we know about the physical characteristics of the Earth's interior comes from the study of how seismic waves, generated by earthquakes, pass through the Earth. Two important types of seismic waves are primary (P-waves) and secondary (S-waves). Primary waves are compressional and pass through materials by jiggling molecules back and forth, parallel to the direction of travel. If one stretches a Slinky and then firmly taps one end, a wave travels down the spring, moving individual sections to and fro—this is like the compressional motion of a P-wave. An important property of P-waves is that their speed increases as the density of the surrounding material increases, and vice versa. P-waves can pass through both solids and liquids. Shear or secondary waves propagate by deforming a material or shifting the molecules from side to side. S-waves can pass through solids, but not liquids, because true liquids cannot deform. By studying how both P- and S-waves travel through the Earth, scientists can estimate the relative hardness or fluidity of the Earth's internal layers.

Near the surface, the crust and upper mantle together form a rigid, hard layer called the lithosphere (from the Greek word *lithos*, meaning stone). In the analogy of a Milky Way bar, the lithosphere is the brittle, thin layer of chocolate on top. The lithosphere extends from the surface to approximately 100 kilometers beneath the oceans and 100 to 200 kilometers below the continents (refer to *Figure 1.3*). Below the lithosphere, a relatively thin zone exists in which both P- and S-waves slow down. This low-velocity layer, approximately 100 kilometers thick, is called the asthenosphere (from the Greek word *asthenes*, meaning weak). The slowing of seismic waves in the asthenosphere suggests that it is partially molten or fluidlike, which allows it to deform plastically, similar to tar or asphalt. In the Milky Way model, the asthenosphere is equivalent to the underlying caramel layer. Below the asthenosphere, scientists think the mantle hardens, but its exact nature remains uncertain. Seismic discontinuities, or changes in seismic wave speed, occur at depths of 410 and 670 kilometers within the mantle and are believed to reflect changes in mineral structure (not composition). New research indicates that at the base of the mantle, the core-mantle boundary, a 5- to 50-kilometer-thick layer exists in which seismic velocities are also reduced. Scientists now believe that a partial melt or fluidlike molten mass may exist at the core-mantle boundary—a surprising development, suggesting a more complex region than previously theorized.

Research also indicates that S-waves cannot pass through the outer part of the Earth's metallic core, which seems to indicate that the outer portion of the core is liquid and the inner solid. The Earth's magnetic field is thought to derive from the planet's spinning about its axis and the subsequent motions of the outer, metallic, liquid core. The transition from liquid to solid at the outer-inner core boundary probably results from increases in both temperature and pressure. On the basis of new seismic data, some scientists believe that the inner core is not a simple metallic solid as once speculated, but may be rotating independently of the planet, may be softer, and may contain iron crystals that are unevenly distributed or oriented throughout the core (Monastersky). These new findings are controversial; further research may or may not lead to confirmation. But undoubtedly as technology improves and scientists continue to study the Earth, we will learn more about the interior and very center of our planet.

Angry Borders

The Earth's surface is divided into about fifteen lithospheric plates that are internally rigid and overlie the more mobile asthenosphere (*Figure 1.4*). The plates are irregular in shape, vary in size, and move relative to one another over the Earth's spherical surface. A single plate can contain oceanic crust,

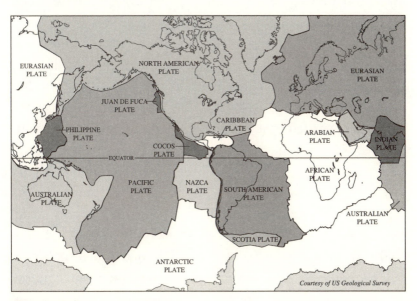

Figure 1.4
The Earth's surface divided into its tectonic or lithospheric plates.

continental crust, or both. They are continually in motion, in relation both to each other and to the Earth's rotation. At plate boundaries, where they constantly jostle and grind against one another, huge mountain chains are built, land is created or destroyed, and deep oceanic trenches are formed. It is also here, at the plates' angry borders, that the Earth's fury is at its peak, creating the majority of the world's earthquakes, volcanoes, and tsunamis. Where plates meet, one of three restless border types is formed: a divergent, convergent, or transform boundary.

Divergent Boundaries

At a divergent boundary, two lithospheric plates are moving away from one another. The mid-ocean ridge system, the most extensive mountain chain on Earth, is a consequence of plate divergence. At the crest of a mid-ocean ridge, lithospheric plates move apart and hot molten rock from deep within the Earth wells upward. While beneath the Earth's surface, hot molten rock is called magma; when it erupts or oozes above ground, it is called lava. Magma is generally a mixture of melted or crystallized minerals and dissolved gases. It is typically less dense than the surrounding materials and is driven upward by buoyancy. At a mid-ocean ridge, magma rises towards the surface and erupts onto the sea floor to create new ocean crust. When hot lava comes into contact with the cold seawater (approximately 2° C, 35° F) at depth, it cools very quickly. This process creates large, rounded rocks called pillow basalts (a dark, fine-grained silicate rock, rich in magnesium, calcium, and iron [*Figure 1.5*] that look something like hardened black toothpaste squeezed from a tube. As Hess and Dietz

Figure 1.5
Underwater photograph of pillow basalts on the Hawaiian
submarine volcano Loihi. Courtesy of the U.S. Geological Survey.

suggested, the separation of plates at a mid-ocean ridge and the
creation of new oceanic crust are now known together as sea-
floor spreading.

Sea-floor spreading occurs intermittently and at varying
rates. In the Pacific Ocean, along the East Pacific Rise, new sea
floor is created at a rate of approximately 6 to 17 centimeters
per year. In contrast, in the Atlantic, along the Mid-Atlantic
Ridge, the rate of spreading is slower, at an estimated 1 to 3
centimeters per year. Variations in heat flow, the chemical
composition of upwelled magma, and the structure of a ridge
along its axis appear to be related to the spreading rate. At the
Mid-Atlantic Ridge, a slow-spreading ridge, the magma is
blocky and relatively viscous, forming a steep, rocky terrain

with a topographic low or valley along the rift axis (color plate 22). At the East Pacific Rise, a fast-spreading ridge, molten material is thinner and less viscous, forming a flat, broad ridge with a topographic high at its center. Scientists speculate that beneath a fast-spreading ridge there exists a narrow zone of high heat and melting. Seismic evidence and three-dimensional imaging suggest that 1 to 2 kilometers beneath the East Pacific Rise lies a thin horizontal layer of molten material that feeds the spreading center. At slow-spreading ridges the axis appears to be cooler, thicker, and subject to greater faulting and earthquake activity.

In Iceland, the Mid-Atlantic Ridge runs right through the middle of the continent. Consequently, in Iceland scientists are afforded an unprecedented look at the processes of rifting along a slow-spreading mid-ocean ridge. Studies indicate that rifting occurs by a slow widening and sinking at the ridge axis, until a breaking point is reached and fractures occur. Cracks begin to form parallel to the rift, earthquakes jolt the region, and lava erupts through some of the fissures (Decker and Decker). Rifting events may be tens of kilometers long, and they tend to occur infrequently at any one spot. Along all mid-ocean ridges, large cross-cutting fractures or transform faults develop owing to the motion of the rigid plates over the underlying spherical surface.

In 1977, while exploring the Galapagos Rift off the coast of Ecuador in the submersible *Alvin* (*color plate 5a*), geologists made a startling discovery. They found chimneylike structures spewing dark clouds of superheated, mineral-rich water (*color plate 5b*). But even more surprising was the presence of abundant and unusual marine life. Giant tubeworms, huge clams, blind shrimp, and crabs were found thriving around the active sea floor vents (*color plate 5c*). At first, biologists listening to the accounts of the dramatic events unfolding beneath the sea did not believe the geologists' strange tale, but later as photographs and data poured in, they were con-vinced—and amazed. Deep-sea vents and their associated communities of strange marine life have now been found throughout the world's oceans. The discovery of this previ-ously unimagined ecosystem thriving in the darkest reaches of the sea was one of the most important and stunning scien-tific findings of the century.

At deep-sea vents or chimneys, plumes of mineral-rich, superheated water erupt from fissures in the sea floor. Heated by molten material below, the water emanating from an active hydrothermal vent field can range between a warm 25° Centigrade (77° F) to a fiery 400° C (752° F). The intense pres-sure at vent depths, some 2,500 meters below the sea surface, allows the water temperature to rise above its boiling point and remain as a liquid—hence the term *superheated*. When the hot vent water mixes with cold seawater at the ocean's floor, dissolved minerals rapidly precipitate out of the "smok-ing" vent water and accumulate rapidly on the surrounding sea floor. At some sites, the continual deposition of minerals from the vents creates huge metal-sulfide deposits called chimneys that grow to towering heights above the sea floor. At hot chimneys (270 to 400° C), the water looks black from min-eral precipitation; these are called black-smokers. At cooler sites (25 to 270° C), water gushing from the vents appears white in color, and these are called white-smokers. One chim-ney deposit on the Juan de Fuca ridge, off the west coast of

Seattle, reaches higher than fifteen stories and was dubbed
Godzilla by the scientists who discovered it (Broad). Off Papua
New Guinea, industry and government officials are considering the mining potential of newly discovered mineral-rich
vent deposits. And in 1998, using a remotely operated vehicle,
scientists were able for the first time to collect large chimney
structures from deep beneath the sea. Intense research on the
recovered samples is now underway and will undoubtedly
reveal much about deep-sea vents and their mineral accumu-
lations and biologic communities.

Scientists are also studying the chemistry and flow of the
water at deep-sea vents. At the mid-ocean ridges, cooling and
contraction of molten rock generates cracks in the underlying
basalt. Cold seawater percolates down through these fissures
and is heated and chemically altered by the hot magma below.
As the seawater heats, its density decreases, and buoyancy drives it back up and through other cracks and holes in the sea
floor. In some areas, superheated water escaping from the vents
becomes trapped under ledges. Because the hot vent water is
less dense than the surrounding cold seawater, buoyancy continues to drive it upward. In video footage, superheated water
slowly escaping from beneath a deep-sea ledge gives the
appearance of a shimmering upside-down waterfall. For
decades, scientists have been puzzled by certain aspects of the
ocean's chemistry. For instance, vast expanses of the sea floor
are covered with manganese-rich nodules, but seawater contains very little manganese. Where does the manganese in the
nodules come from? Manganese in the outflow of deep-sea
vents may provide the answer to this and other chemical mysteries in the sea. The process by which seawater flows in,

through, and out of deep-sea vents is known as hydrothermal circulation and is now believed to play a major role in the ocean's chemistry and heat flow.

The discovery of new, bizarre forms of marine life at deep-sea vents was undoubtedly one of the most spectacular events in biological oceanography in the 20th century. More than 300 new marine species have been found, including new types of clams, tubeworms, shrimp, crabs, mussels, and pink sea urchins. Unlike all other ecosystems on Earth, deep-sea vent communities do not derive their energy from the sun through photosynthesis, but are chemosynthetic, using the chemical energy in sulfide compounds from vent water. Bacteria, found in overwhelming abundance at the vents, essentially feed on (oxidize) sulfides in the hot chimney plumes. Some bacteria provide nutrition for vent organisms, such as tubeworms or clams, by actually living inside the organism. Other bacteria are free-living and are eaten by various members of the vent community. Because these bacteria live in such hot, hostile, and extreme conditions, they have been called extremophiles or thermophiles. New research suggests that some of the bacteria may not actually be bacteria, but a very old and newly discovered form of microbial life dubbed Archea. The abundance, diversity, and growth of life at active vent sites are greater than was ever expected in the deep sea. And, some scientists now believe that life on Earth may have originated in vent-type conditions. Today, we continue to study sea-floor spreading and deep-sea vents to answer some of the many persisting questions. What controls the rate of sea-floor spreading? Why are some areas hydrothermally active and others dormant? How deep into the Earth does seawater circulate, and how far down

Oceanic-Oceanic Collisions

Off the coast of the Philippine Islands in the Pacific Ocean lies the deepest site in the sea, the Marianas Trench, some 11 kilometers (7 miles) deep (at 35,838 feet, this trench is deeper than Mount Everest is high). Beneath the Marianas Trench two plates of oceanic crust are colliding, the Pacific Plate and the Philippine Plate. Typically, when two oceanic plates converge, the older, denser plate is driven beneath the younger, less dense plate (as ocean crust ages and spreads away from a mid-ocean ridge, it cools and its density increases). However, exceptions do inexplicably occur. For instance, in the Caribbean the younger Caribbean Plate is being driven beneath the older South American Plate. When one plate is forced beneath the other, the process is called subduction, and the area in which this occurs is called a subduction zone (*Figure 1.6, color plate 6*).

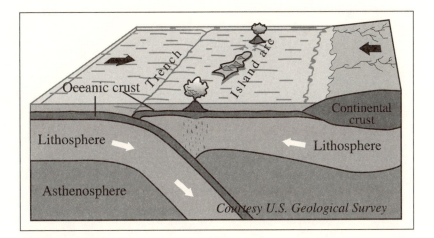

Figure 1.6
The collision of two tectonic plates, each with oceanic crust on its leading edge.

Ocean trenches, like the Marianas Trench, are the surface expression of a subduction zone. Water deep within the Earth is now thought to be an important lubricating agent in the subduction process, allowing one plate to slide over another. Faulting within a subduction zone can produce some of the largest and most devastating earthquakes on the planet. In turn, these earthquakes and movements of the sea floor can generate very long waves that come ashore as towering tsunamis. Additionally, high temperature deep in a subduction zone melts the downgoing slab and generates molten rock. Driven by buoyancy, the hot magma flows upward through fractures and pores in the overlying rock and erupts at the surface to form a chain or arc of active volcanoes behind the subduction zone. An arc of volcanoes known as the Ring of Fire rims the Pacific Ocean (*Figure 1.7*). Seventy-five percent of the Earth's active volcanoes and most of the planet's earthquakes and tsunamis occur as a result of the Pacific's infamous Ring of Fire.

Oceanic-Continental Collisions

Oceanic crust is denser than continental crust, so when the two collide, oceanic crust is forced downward, beneath continental crust. For instance, at the Peru-Chile trench the oceanic Nazca Plate is being driven beneath the South American continent, part of the South American Plate. Behind the subduction zone, great upheavals of the land and slow continuous uplift have created the lofty Andes Mountains. During collisions of oceanic and continental crust, or oceanic and oceanic crust, some of the sediment and rock on the downgoing slab may be scraped off and accreted onto the overriding plate. The island of Barbados is built on a wedge of material scraped off the

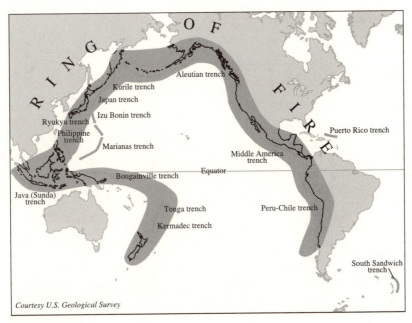

Courtesy U.S. Geological Survey

Figure 1.7
The infamous Pacific Ring of Fire.

Caribbean plate as it dives beneath the South American plate. Subduction at oceanic-continental plate collisions also creates and sustains many of world's most active volcanoes, causes strong earthquakes, and triggers tsunamis.

Transform Faults

Transform faults are where two plates slide in opposite directions past each other. The large fractures that cut across the mid-ocean ridges are considered transform faults. Shallow earthquakes occur frequently along transform faults. The most famous transform fault is the San Andreas, where the Pacific Plate moves northwest against the southeasterly moving North American Plate (more on this in the next chapter).

Hot Spots

The majority of volcanoes, earthquakes, and tsunamis occur at plate borders, but we occasionally find volcano formation, faulting, and associated seismic activity within the interior of a lithospheric plate. These volcanoes are derived from a localized source of heat or "hot spot" beneath the lithosphere. Much like a conveyor belt, a tectonic plate moves over the Earth's surface and may pass over a stationary hot spot or thermal plume. The rising heat and erupting magma generate a series of seamounts or volcanic islands that trace the movement of the plate over the hot spot. The Hawaiian Island chain is the best-known product of a hot spot (*Figure 1.8*). As the Pacific Plate moves over an underlying hot spot, the Hawaiian Islands are being created, Hawaii being one of the most recent to pass over. By dating rocks on the islands, scientists have determined that the Pacific Plate has moved an average of 8.6 centimeters per year for at least 70 million years. A bend in the island chain suggests that some 40 million years ago, the movement of the Pacific Plate changed direction, from north to northwest.

Scientists now believe that hot spots may be the dominant means by which oceanic crust is modified between creation at mid-ocean ridges and subduction in deep-sea trenches. In addition, because the lava at hot-spot volcanoes is chemically different from that at mid-ocean ridges, hot spots are thought to originate from thermal plumes deep in the mantle, below the asthenosphere and possibly even at the core-mantle boundary. No obvious pattern in the distribution of presently active hot spots has been determined, although some concentration seems to occur in regions away from subduction zones. Hot spots occur less commonly under the continents. The famous

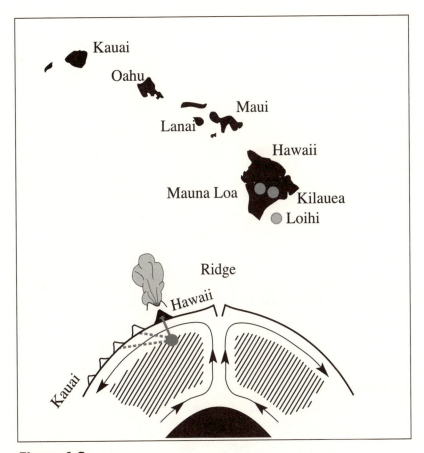

Figure 1.8
Hawaiian islands and the underlying hot spot.

geysers, boiling mud pools, and steaming landscapes of Yellowstone National Park are thought to result from an underlying continental hot spot. Research also suggests that some 100 million years ago, hot-spot activity was five to ten times greater than it is today, and may have had a significant impact on ancient ocean temperatures, sea level, and climate. Many puzzling questions about hot spots remain unanswered.

Scientists are actively investigating the source of hot-spot heat and magma, the causes of change in hot-spot activity, the controls on their spatial distribution, and links to climate.

The Force

Ever since the early continental drift and plate tectonic theories were proposed, scientists have struggled to understand why and how lithospheric plates move. Even today our understanding of the forces that drive plate motion is rudimentary. Much of the difficulty arises from the fact that we have no method to view directly the planet's interior or test relevant theories. None of the presently proposed mechanisms of plate motion seem to explain all aspects of plate movement, but for now, they are our best guess based on the available evidence.

Plate motion appears to be driven mainly by convection within the Earth's mantle layer and by pull from plate subduction (*Figure 1.9*). As discussed earlier, the asthenosphere, a thin layer in the upper mantle, is believed to be partially molten. Heat from deep within the Earth is thought to cause very slow convection currents in the partially molten or fluidlike asthenosphere (or possibly below it). Convection, a means of heat transfer, occurs when a temperature increase causes a liquid or gas to expand, become less dense, and rise within the surrounding material. As the substance rises, heat is transferred to the surrounding material. The colorful motion in a lava lamp is a product of convection. At the lamp's base, heat from a light warms colored wax in a surrounding oil mixture. As the wax heats up, it becomes less dense than the nearby oil and rises. Near the top of the lamp, away from the

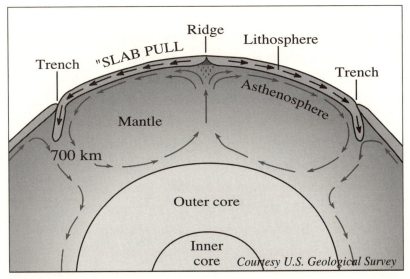

Figure 1.9
Convection and slab pull in the Earth's mantle and
asthenosphere.

source of heat, the wax cools, becomes more dense, and sinks
back to the bottom of the lamp.

The heat source for convection within the asthenosphere
comes from deep within the Earth's interior, fueled by the
decay of naturally radioactive materials (e.g. uranium, plutoni-
um, and thorium) and heat from the early formation of the
planet. In a simplistic, conceptual model of what must be a
spatially complex and dynamic system, we envision a series of
convection cells within the asthenosphere. Uneven heating
causes thermal plumes to rise at mid-ocean ridges and cooling
near the surface creates descending plumes at subduction
zones. In between, the asthenosphere is thought to move horizontally from beneath a spreading center, a ridge, to a subduction zone, a trench. Friction between the lithosphere and the

underlying asthenosphere acts like glue, and the lithospheric plates are dragged along by the motion of the asthenosphere. To envision how the Earth's rocky mantle could flow almost like a fluid, think of the child's toy Silly Putty. When molded tightly into a ball, a large mass of the squishy stuff will bounce like a solid off a hard surface. But if placed in one spot for an hour or more, the Silly Putty, like a very thick fluid, will gradually spread out over the surface. It acts like a solid on a short time scale, and a fluid over longer periods. Mantle convection may be similar, flowing very slowly over a very long period.

At subduction zones, gravity pulls the slabs of cold, dense oceanic crust down into the mantle. This process is now believed to play a major role in plate motion, as it results in the entire plate being pulled along with the descending slab. In the past, scientists also speculated that the formation of new crust at the spreading ridges pushed the plates apart. Now most believe that the horizontal force created during spreading is minor compared to the "slab pull" at the subduction zones or convection within the asthenosphere. The source of plate motion will undoubtedly be the focus of scientific study and debate well into the future.

The Evolving Surface

For hundreds of millions of years, plate tectonics has been at work, and the Earth's surface has been evolving and changing shape. Continents have collided to form mountains, oceans have been born, and large sections of the planet's crust have been obliterated in subduction zones. Because of the recycling of the Earth's outer skin, the oldest ocean crust is only 180 million years old. This makes it difficult to reconstruct what the planet's surface looked like earlier than about 200 million years ago. Scientists must rely on rocky clues from the much older continents. For instance, the Appalachian Mountains along the East Coast of North America are thought to represent the suture zone of a powerful plate collision that occurred some 500 million years ago.

Scientists now speculate, reviving Wedener's once-rejected theory, that relatively early in the Earth's history, some 600 million years ago, the continents were joined as one large land mass, aligned along the equator. Over time and through the movement of the tectonic plates, the continents split apart and shifted. However, by about 225 million years ago, the continents had once again come together to form another supercontinent, which geologists have named Pangea, and one universal ocean, known as Panthalassa (*Figure 1.10*). Evidence suggests that by about 200 million years ago, the supercontinent had again been ripped apart, this time forming two major continents and a smaller sea, the Tethys. By 135 million years ago, India, Antarctica, and Australia had broken away from the other continents, and a major restructuring of the Earth's surface took place. During the next 70 million years, the Atlantic Ocean formed, India moved menacingly northward toward Asia, and Africa rotated counterclockwise. Eventually, India collided with Asia, Australia moved northward, North and South America became attached, and the Atlantic and Indian Oceans widened. Today, we see a distribution of the continents and ocean that reflects a very busy past. However, what we see today is by no means the conclusion to this long-drawn-out tale. Through sea floor spreading, the Atlantic and Indian

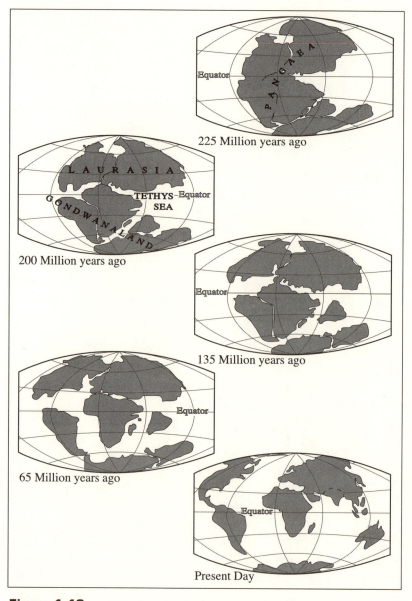

225 Million years ago

200 Million years ago

135 Million years ago

65 Million years ago

Present Day

Figure 1.10
The approximate position of the continents through time, 225
million years ago until the present. Courtesy of the U.S.
Geological Survey.

Oceans continue to widen, while the Pacific Ocean shrinks. Africa is being split in two at the East African Rift, and the Mediterranean is getting smaller. As time moves on, so will the Earth's tectonic plates. Their motion will continue to shape and change the planet's surface, and the jostling and grinding of their edges will inevitably produce powerful volcanic eruptions, cause strong earthquakes, and trigger towering tsunamis.

Chapter 2

Earthquakes

Northridge, California, 1994

At 4:31 on the morning of January 17, 1994, everyone within greater metropolitan Los Angeles was wide awake and focused on one thing-the violent shaking of the ground. For many, the next ten to fifteen seconds were a struggle between life and death. For most, it was an adrenaline overload that they would rather have done without. It was a ground-shaking, glass-shattering event that became known as the Northridge earthquake. In just fifteen seconds or so, the quake caused between $20 to $40 billion in property damage, fifty-seven deaths, 11,000 injuries, and left more than 20,000 people without homes. The event also raised many questions about the safety of manmade structures, including some of the region's freeway bridges, which collapsed like concrete pancakes (*Figure 2.1*). To add insult to injury, during the following week more than 745 aftershocks, large enough to feel, rumbled through the area. By the end of one month, a total of 930 palpable tremors had occurred. Residents of the San Fernando and Simi Valleys were left shaken, scared, and sleepless for weeks. Five years later, the

Figure 2.1
Freeway bridge collapse in California during the Northridge
Earthquake. Courtesy of the U.S. Geological Survey.

number of aftershocks had climbed to an incredible total of 14,849 (including those too small to feel, but large enough for instruments to detect).

On the Richter scale, the Northridge event registered a magnitude of 6.7. Earthquakes of a magnitude of 2.0 or less are relatively weak and may not even be noticed. Those above about 7.0 are considered strong, and when an earthquake reaches a magnitude of 8.0 or more, it is known as a great quake. The Northridge earthquake was a typical magnitude 6.7 event, but unfortunately it distinguished itself by occurring directly beneath a major metropolitan center. Why didn't scientists predict the Northridge quake? Is it common for so many aftershocks to occur? And can we prevent catastrophic damage, such as freeway collapses, when earthquakes strike? These questions and others like them are what scientists who study earthquakes, and the damage they cause, seek to answer.

Worldwide, there are two or three magnitude 6.5 or larger quakes every week. In California, earthquakes of this magnitude occur, on average, about once every six years. Luckily, most earthquakes are small, and many happen in unpopulated regions, causing little damage and few injuries. In California, thousands of earthquakes occur every year; most are too small to feel. As a result, California has become a mecca for earthquake research, and much of what we know about earthquakes and preventing earthquake damage has come from studies within the region.

Surprisingly, though, California is not the state with the most earthquakes in the United States; that dubious honor goes to Alaska. In fact, both the total number and size of the largest earthquakes in California are significantly less than in many other parts of the world, including the Far East and the west coast of South America. But California probably does have the world's safest buildings and is the best prepared for earthquakes. Tragically, history provides many examples of what can happen when a quake strikes an area ill prepared for the Earth's fury. On February 29, 1960, a moderate-sized quake, about magnitude 5.9, rocked Agadir, Morocco. Of the 30,000 residents of Agadir, approximately one-third were killed and another third injured. The old masonry core of the city was completely obliterated. Another moderate earthquake, also magnitude 5.9, causing heavy damage and loss of life, occurred just recently in Colombia, on January 25, 1999. Here again, unreinforced stone and adobe construction collapsed, while more modern housing stood unscathed. Because the devastated area was a mountainous one, damage to roads hampered the delivery of relief supplies. In comparison, a quake of almost exactly the same magnitude struck near Whittier Narrows, California, east of downtown Los Angeles, on October 1, 1987. The Whittier Narrows quake caused an estimated $358 million in property damage, but only eight people were killed (some by heart attacks) and about two hundred injured. By studying how structures behave in an earthquake, scientists and engineers have steadily improved the integrity of construction during earth-shaking events.

Earthquakes in California or anywhere else in the world clearly can have catastrophic consequences, but each event provides scientists with important information about how and why earthquakes happen, which may help reduce deaths, injuries, and property damage in the future.

Studying Earthquakes

Most of what we know today about earthquakes and the interior of our planet comes from seismology, the study of seismic waves, vibrations that pass through the Earth. An earthquake is a seismic vibration caused by the breaking or slipping of rocks somewhere in the Earth's crust. Seismic waves are detected using a device called a seismometer and recorded by a seismograph. Seismographs are operated from seismic stations, and the recordings made by a seismograph are called, not surprisingly, seismograms (*Figure 2.2*).

Although seismology is a rather young science, earthquakes and their devastating damage have been documented

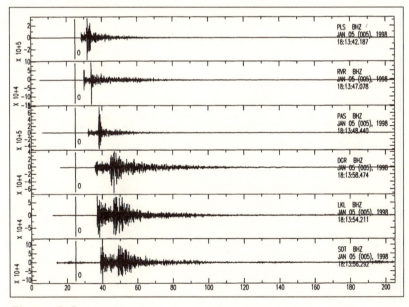

Figure 2.2
Seismograms from six seismic stations. Time increases to the right, and the vertical scale indicates the strength of the ground motion.

throughout human history. Some of the earliest written records tell of the Earth's great rumblings and subsequent destruction. Evidence of even earlier earthquake events comes from archeology. For example, an earthquake appears to have caused extensive damage to the ancient Greco-Roman city of Kourion on Cyprus. Coins found in the ruins put the date of the destruction at around 365 AD.

The systematic collection of earthquake data using instruments specifically designed for the task began sometime in the mid- to late 1800s, and by the 1890s, early seismographs were in fairly wide use in both Europe and Japan. The first seismographs were crude by modern standards, but they worked fairly well. In fact, some of the early seismograph designs remained in use, almost unchanged, until the 1980s.

A seismograph or seismometer is designed to record or measure the ground's movement during an earthquake. Ideally, a seismograph would measure the changing distance between a fixed point and the moving ground. Unable to create such a device, scientists early on devised an instrument based on the principles of a pendulum. In a seismograph, a pendulum weight is hung loosely from a frame that is firmly attached to the Earth. When the Earth moves either horizontally or vertically, the pendulum tends to stay still, at least momentarily, and then begins to move. The stronger the shaking, the larger the relative motion between the pendulum weight and the frame. To get an accurate reading for every earthquake, the pendulum must be slowed or stopped after each event. During an event, the record of a pendulum's position over time looks like a zigzagging wave, slowly decreasing in size after the ground has stopped shaking. (refer to *Figure 2.2*)

The principles of a seismograph are actually fairly simple; the challenging part was to design a means of measuring the position of the pendulum with respect to the frame. Some of the earlier seismographs were entirely mechanical. The pendulum was linked to a pin or stylus that scratched a line in a layer of black soot on a piece of rolling paper. One of these early instruments, the Wiechart seismograph operated at Potsdam in Germany around the turn of the century, and weighed an ungainly 21 tons. Clearly not a portable model. To overcome the friction between its own mechanical parts, the early seismographs needed to be large. More modern designs use an electromagnetic linkage. The pendulum carries a tightly wound coil of wire that hangs between the poles of a magnet. When it moves, an electric current is generated in the coil, and this produces an electrical mimic of the ground motion. The small electric current can then be used to drive a stylus, move a beam of light that registers on photographic paper, or drive an ink pen that records on regular paper. It can also, as is currently the standard, be recorded on a magnetic or optical disk by a computer. Modern seismographs and seismometers are much smaller and more precise than the early versions, and some are even portable.

Seismographs and seismometers can be remarkably sensitive and easily detect earthquakes that are far too small for a person to feel. They can also record the weak ground motion of large, distant earthquakes. An earthquake that registers 6.0 or higher on the Richter scale can be detected by seismometers anywhere in the world. Most seismographs are so sensitive they can detect not only distant earthquakes, but also the not-so-distant movement of trains and trucks, high surf, the impacts of meteorites, dynamite blasts at mines and quarries, underground nuclear test explosions, and even the wind blowing on nearby trees and buildings.

One of the first exciting discoveries made by seismologists, which now seems rather obvious, was that earthquake motions take time to travel through the Earth. Earthquakes generate shock or seismic waves that travel through the crust, like the waves that radiate outward from a stone thrown into a pond. The detectable motion arrives quickly at seismic stations close to the quake, and later at stations farther away. People who happen to be on the telephone during a quake report that a time delay can easily be noticed; the person closest to the quake's origin feels it first.

Another early discovery was that each seismic record shows more than one burst of activity. These are now known as an earthquake's "arrivals" or "phases." Multiple phases occur partly because, as discussed in the first chapter, in a solid like the Earth's crust, different types of seismic waves are generated. The fastest, and hence the first to be recorded, are the compressional primary or P-waves. The secondary or S-waves propagate more slowly and so are detected a bit later. Both types of seismic waves travel at different speeds through different types of rock. For most seismically active areas, these speeds are well known. Hence, the arrival times of P- and S-waves at various stations can be used to determine the location of an earthquake. Actually, an observer with sufficient presence of mind can sometimes feel both P- and S-waves separately and, noting the time delay between them, guesstimate how away the source of the earth's shaking is. It is not uncommon to find a seismologist looking at his or her watch while the ground is moving. Earthquakes also generate another type of wave that propagates even more slowly. Collectively, these are known as surface waves, because they always move on the surface and do not travel down into the Earth, like P- and S- waves do (P and S waves are also referred to as body waves). Surface waves tend to be felt as a rolling motion, as compared to the sharp thud or jolt of a compressional P-wave.

Another classic and important means of studying earthquakes is to map and monitor the actual movement of the earth's crust or plates leading up to and during a seismic event. In the early 1900s, making accurate measurements for mapping and monitoring of plate motion was much more difficult than it is today. Traditional surveying techniques involve a method called triangulation. An instrument known as a theodelite is used to measure the angles between various survey markers, and a level is used to measure differences in elevation. When these measurements, along with a measure of distance, are made repeatedly at the same place over time, they reveal the movement of the Earth's tectonic plates.

Today, surveying and monitoring techniques often rely on laser technology and the Global Positioning System (GPS), originally a navigation system for the U.S. military, to make faster and more precise measurements. Each satellite within the GPS broadcasts a signal specifying its position at any given time. On the Earth's surface, a GPS receiver calculates the distance to the satellite based on the transit time of the signal. With one or a few receivers, the accuracy of this method is within meters; with a network of receiving stations, a site can be located within a few centimeters. When the location of fixed stations on the Earth's plates is repeatedly measured relative to the orbiting grid of GPS satellites, plate movement over a given period can be precisely determined. This method has almost completely replaced traditional surveying and triangulation techniques. Satellite imagery is also providing a space-borne view of the Earth's surface, and can be used to help define large-scale fault patterns.

Today's technology can go only so far, however, when it comes to determining a region's history of earthquakes and plate motion. In some densely populated places like California, data from triangulation measurements over a long period are available and can be compared to reveal the region's seismic history and motion on specific faults. In other areas, historical accounts of earthquakes, often from meager newspaper reports, must be used to estimate quake frequency, size, and fault offset. For a longer-term perspective, scientists must often look to a region's geology for clues to plate motion. For example, along the San Andreas Fault in the Carrizo Plain (*color plate 7*), between Los Angeles and San Francisco, some of the stream beds have been offset or bent by one or more earthquakes (*Figure 2.3*). If such formations and their changes can be dated, the rate of plate motion can be deduced.

Modern technology is improving exponentially with the development of new techniques, sensors, and instruments to monitor and study earthquakes. Scientists today also use measurements of the Earth's magnetic field from aircraft to locate faults, and through computer models, they can simulate earthquakes to study how they deform the land, create shaking, and release energy.

Figure 2.3
Aerial photograph of a stream offset in California. Courtesy
U.S. Geological Survey.

X Marks the Quake: Locating the Origin

The first seismic wave to arrive at a seismic station represents
the initial pulse of an earthquake's motion, and it is used to
determine where the earthquake actually began. More wave
motions arrive in the next few seconds, but the initial pulse

is the most clearly defined and easily measured. The exact location within the Earth where an earthquake begins is called the hypocenter or focus. Above the hypocenter is a spot on the Earth's surface called the epicenter (*Figure 2.4*). The epicenter may or may not be the center or area of most damage, because the hypocenter is only where a rupture or tear on a fault first starts. A fault is simply a fracture in the Earth's crust along which breakage and displacement occurs, often called offset. Usually, but not always, the breakage and displacement of a fault happens quite suddenly and causes vibrations or seismic waves to radiate outward, which we feel as ground shaking. A break or rupture can spread a significant distance along the fault away from its starting point, or the hypocenter.

Scientists use the arrival times of P- and S-waves at several seismic stations to precisely locate an earthquake's epicenter

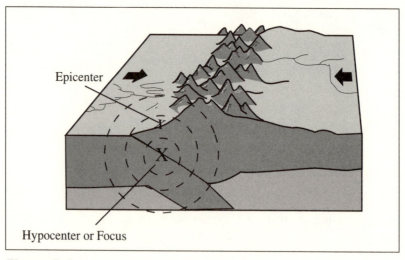

Figure 2.4
An earthquake's epicenter and hypocenter.

and hypocenter. From one station, it is only possible to determine how far away the epicenter was from that individual station. The earthquake could have originated anywhere on a circle having a radius of that distance, with the station at its center (*Figure 2.5A*). With two stations, two possible epicenters produce the right arrival times: where two circles, one drawn around each station, intersect (*Figure 2.5B*). Data from a third station distinguishes between these two possible locations (*Figure 2.5C*), and a fourth station adds enough information to determine the quake's depth within the Earth, its hypocenter. Additional data allow for even more precision. Thus, having a network of seismic stations is essential for pinpointing the source of any earthquake. Precise timing is also needed to compare seismic recordings from different stations. Clock accuracy is not a problem now, but in the early days of seismology, when clocks were generally pendulum-driven and "drifted" by minutes per day, timing could be a major chore when trying to locate an earthquake's origin.

Going beyond detecting earthquakes, the study of seismology can tell us much about the interior of the Earth. P- and S-waves, as opposed to surface waves, travel not only horizontally to nearby seismic stations, but also downward, deep into the Earth. Often seismic stations on one side of the globe can detect large earthquakes that originated on the other side. A P-wave, the fastest type of wave, takes about 20 minutes to travel straight through the Earth. Each time a seismic wave encounters a boundary within the Earth's internal layering, part of the energy passes over the boundary, perhaps bent slightly like light through a prism, and part of the energy is reflected back. Seismic stations all over the world record

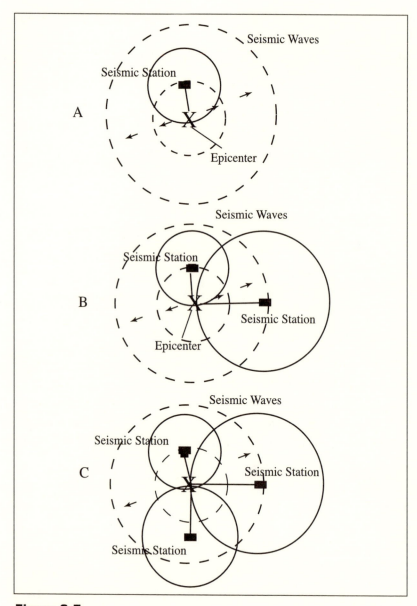

Figure 2.5
Locating an earthquake's origin from (A) one, (B) two, and
(C) three seismic stations.

these refracted and reflected phases, and with experience, seismologists can use the data to distinguish the Earth's internal layers. We could say that earthquakes have allowed us to "x-ray" the inside of the Earth with seismic waves. With lots of data in a specific region, modern computer technology now allows us to "CAT scan" or image the inside of the Earth using seismic waves.

In the early days of seismology, the emphasis was on recording large earthquakes from all over the world. Data were collected and compiled into bulletins of phase arrival times, which were exchanged between stations by mail, sometimes even across hostile boundaries. The sharing of information on earthquakes took precedence over political ill will.

Where Earthquakes Occur

When the global distribution of earthquake epicenters is plotted on a map, a striking pattern is revealed, one that is neither uniform nor random (*color plate 4*). The geographic and time distribution of earthquakes is referred to as seismicity. Clearly, some parts of the Earth exhibit greater seismicity than others. A large percentage of the world's earthquakes occur in one arching band, extending around the rim of the Pacific Ocean—the notorious Ring of Fire. Another distinct band of seismic activity extends through the Middle East and southern Europe. When first examined, some of the patterns in seismicity were old news; the Ring of Fire, the Middle East, and southern Europe were already famous for devastating earthquakes. But the global distribution of seismicity also revealed another remarkable and unexpected pattern-secondary bands of seismic activity occurred under the oceans along the features now known as the mid-ocean ridges. In fact, the worldwide seismicity map looks almost like a jigsaw puzzle, with the earthquakes outlining the puzzle pieces. Today we call these pieces tectonic plates, and the distribution of earthquakes was one of the major clues that led to the development of the plate tectonics theory in the 1960s and 1970s (chapter 1).

The key to the relationship between tectonic plates and earthquakes is movement. For example, a line of earthquakes marks the Mid-Atlantic Ridge, the huge undersea mountain range that meanders down the center of the Atlantic Ocean. At the ridge, two tectonic plates are moving apart, widening the Atlantic Ocean and causing North and South America to become progressively farther away from Europe and Africa. However, earthquakes occur more frequently and are the most severe not where tectonic plates are moving apart, but where they are converging.

The Earth's convergent boundaries, particularly where subduction zones occur, can generate earthquakes deep in the Earth, sometimes as far down as several hundred kilometers. Most earthquakes, however, including those in subduction zones, are shallow and occur in the Earth's upper 32 kilometers (20 miles). Deep-focus earthquakes puzzled scientists when they were first observed. Since the Earth's mantle is much hotter than its crust, scientists assumed that as a slab of rock descended, it would become too soft and pliable for faulting and earthquakes to occur. The mantle's consistency is something like a piece of taffy; when frozen it becomes brittle and can be snapped in two, but when warmed the taffy is soft and pliable. Research later revealed that earthquakes occur deep in a subduction zone because a descending slab may remain relatively cool and brittle to a much greater depth than previously thought, nearly 700 kilometers (430 miles).

The Pacific Plate and the Cocos Plate, which together comprise most of the Pacific Ocean floor, are the fastest moving of the Earth's plates, and where these two plates are subducted, the largest and greatest number of earthquakes occur.

Earthquakes also occur at transform faults, the third type of plate boundary. Here, two plates slide horizontally past each other. Quakes at this type of plate boundary are in the minority, but there are a few, very famous examples. One is the San Andreas fault in California (*Figure 2.6*), which marks the division between a small sliver of North America's crust, including Los Angeles, San Diego, and Baja, Mexico, that moves toward the northwest on the Pacific Plate, and the rest of the state, which rides atop the North American Plate, moving toward the southeast. Another example of a plate boundary with predominately horizontal motion is the Dead Sea Rift, which underlies the Dead Sea and Jordan River in the Middle East. Transform faults, like all faults that display horizontal motion, are called strike-slip faults, because the direction of "slip" is parallel to the direction the fault "strikes" across the landscape (*Figure 2.7A*).

In addition to strike-slip faults, there are two other main types of faults, normal and thrust faults. In normal faults the crust on either side of the fault is being pulled apart, and one side slides downward relative to the other (*Figure 2.7B*). Normal faults commonly occur at diverging plate boundaries.

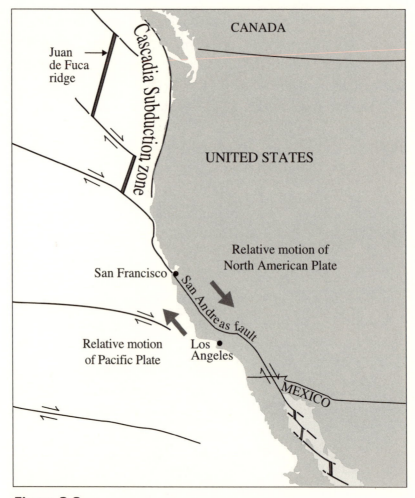

Figure 2.6
The San Andreas fault and Cascadian Subduction Zone.
Courtesy U.S. Geological Survey.

In thrust faults, the crust on both sides of the fault is being compressed, forcing one side of the fault up and over the other (*Figure 2.7C*). Thrust faults are common in subduction

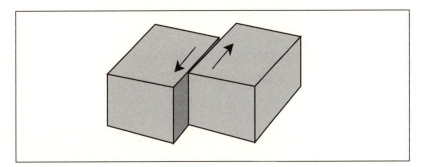

Figure 2.7a
Main fault types: strike-slip.

zones, where the tectonic plates are converging. Blind thrust faults are thrust faults hidden beneath the surface and can cause uplifting of the ground with little evidence of offset at the surface. Both the Northridge and the Whittier Narrows earthquakes occurred on a system of blind thrust faults under Los Angeles.

Earthquakes can also occur, although less commonly, on the interior of a plate. As a plate moves, its interior is stressed, and old faults, perhaps part of an old plate boundary, can be

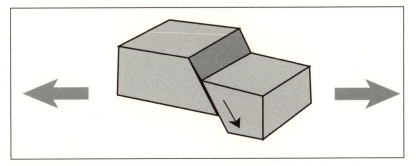

Figure 2.7b
Main fault types: normal.

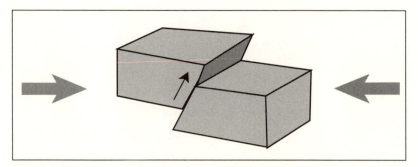

Figure 2.7c
Main fault types: thrust.

activated. Probably the most famous intraplate earthquakes occurred near New Madrid (pronounced MAD-rid by the inhabitants), Missouri, in 1811 and 1812. In total, there were four powerful quakes, each registering a magnitude between 7.5 and 8.0. Because the sediments in the Mississippi Valley are soft, the shaking was extremely strong. Witnesses described the ground as rippling like a field of wheat in a breeze. Surface faulting during the quakes actually altered the course of the Mississippi River, and in some areas dry land was transformed almost instantaneously into a swamp or lake. In addition, because of the nature of the underlying sediment, plumes of quicksand gushed skyward like geysers, reaching as high as 10 meters (30 feet) into the air. The largest quake was felt as far away as Boston and Washington, D.C. Fortunately, the population of New Madrid and the surrounding area in 1812 was sparse, and their log cabins were very resistant to earthquake damage. Current efforts are being made to better assess the potential for future earthquakes in the New Madrid seismic zone and the risks they pose to the now much denser population.

Why Earthquakes Occur

Most earthquakes are caused by the bumping and grinding of tectonic plates along their borders. Volcanic activity can also produce earthquakes. The mechanics of how earthquakes happen was surmised many years before the development of plate tectonic theory. In 1906, one dramatic and destructive event brought scientists a new understanding of how and why earthquakes occur. The San Francisco earthquake happened in a region where there had already been extensive surveying of the land; post-quake measurements, for the first time, revealed important information on the science and nature of earthquakes.

Survey data within the San Francisco area showed that although no actual slip had occurred on any of the faults in the region, for decades prior to the 1906 earthquake, the Earth's crust had been bending. Essentially, crust on one side of town was moving northwest, while on the other side of town, it was moving southeast. In between the two, the crust was being deformed. On the morning of April 18, 1906, the breaking point was reached, and the crust snapped along the San Andreas Fault. Huge blocks of crust on either side of the fault abruptly returned to their original, unstrained shapes, and in the process caused violent waves of shaking in the ground. Suddenly, straight roads, streams, and fences that crossed the San Andreas were offset 3 meters or more. And the furious motion of the earth caused buildings to collapse and ignited fires that spread through and ravaged the city. From triangulation data, geologist Harry Fielding Reid deduced that during the earthquake, strain that had been accumulating for years was released when the deformed crust snapped and elastically rebounded-the land had literally snapped back into place. Reid had no explanation for why strain had been accumulating, but he was able to estimate how fast it was occurring and that it would take, at the observed rate, at least two centuries to accumulate enough to evoke another earthquake of the same magnitude.

Scientists have now constructed a fairly clear picture of how earthquakes happen. Tectonic plates move along at a slow but steady rate. However, along their boundaries the plates do not slide smoothly. Instead, the fault or faults along the boundary between the plates get stuck together or "locked" by friction. As a consequence of this sticking, the surrounding crust bends or deforms. At some point the strain becomes too great, friction is overcome, and a fault or some significant patch of a fault literally snaps (elastic rebound). It is somewhat like stretching a rubber band. The farther it is stretched, the tighter it becomes, until eventually it snaps or breaks apart. An earthquake occurs when the crust snaps and the energy built up and stored within is released as waves of vibration into the surrounding ground. The strength of the earthquake depends on the amount of strain energy released, which is a function of the size of the fault rupture and the resulting offset. After a fault tears, and strain is released, friction again takes hold, and strain begins once more to accumulate.

How Big Is Big?

In general, people tend to think of an earthquake's size in terms of how much the earth shook (its magnitude) and how much damage was caused (its intensity). In the early days of seismology, the only way to quantify an earthquake's size was to measure its observed effects or damage. The most widely used scale to describe an earthquake's damage is the Modified Mercalli Intensity (MMI) scale: twelve points, represented by Roman numerals, ranging from I (1, not felt) to XII (12, total destruction). The scale is meant to cover a range of observable impacts, from almost imperceptible shaking to catastrophic destruction. Low numbers generally describe how people feel an earthquake, and high numbers characterize structural damage. In the town of Northridge and other communities in the northern San Fernando Valley, the peak MMI in the Northridge earthquake was X (10).

In areas prone to earthquakes, it can be instructive to map the damage using contours of similar MMI measurements—much like a topographic map where contours represent similar elevation. In the past, a rough estimate of an earthquake's size could be made based on a chosen contour interval: the magnitude of the quake was proportionate to the size of the area in which it was felt, enclosed on the map by a specific contour. From this mapping, the earthquake's intensity could be gauged, and it was possible to identify areas prone to damage.

It took the advent of widespread seismic recording, however, to make possible a quantitative measure of earthquake size, known as magnitude. Whereas earthquake intensity is a measure of damage and can vary from one location to another, magnitude is intended to be a single number describing the size of an earthquake and does not vary from region to region or place to place.

In the early 1930s, Drs. Charles Richter and Beno Gutenberg, at California Institute of Technology's Seismological

Laboratory, wanted to publish a catalog or list of earthquakes recorded in southern California. However, a list of dates, times, latitudes, and longitudes made little sense without some indication of the relative size of each quake. So Richter, having had some training in astronomy, borrowed the concept of magnitude from stellar astronomy to describe an earthquake's size. In astronomy, *first magnitude* stars are the brightest and *sixth magnitude* stars are the faintest ones that the naked eye can see from a favorable location, like the middle of the desert.

Richter's definition of the "local magnitude" scale, which the press eventually popularized as the Richter scale, is very simple: an earthquake's magnitude is the logarithm to the base 10 of the amplitude on a Wood-Anderson seismometer located 100 kilometers from the epicenter. What does this mean? In the 1930s, every seismic station in southern California was equipped with, among other types, one or two Wood-Anderson torsion seismometers. The Wood-Anderson seismometer was another Caltech invention (by Harry O. Wood and John Anderson) and had many advantages over the other available models. It was small, relatively portable for the time, and most importantly, easy to calibrate. The amplitude is the size of the peak deflection of the pen, stylus, or light beam (in the case of the Wood-Anderson seismometer) produced by a quake on a seismogram. Using the logarithm to the base 10 of the amplitude simply means that for every whole number increase on the magnitude scale, the amplitude goes up by a factor of ten. Hence, if a quake produces a peak deflection of 1.0 millimeters on a Wood-Anderson seismometer 100 kilometers away, the quake has, by definition, a local magnitude of 3.0. If the peak amplitude is 10.0 millimeters, the magnitude is 4.0. If the amplitude is 100.0 millimeters, the magnitude is 5.0, and so on. In other words, an earthquake of magnitude 5.0 has ten times the ground motion of a quake of magnitude 4.0. Rarely is a station located exactly 100 kilometers from an earthquake's epicenter, however, so what the amplitude would have been at that spot must be estimated. To do this, seismologists, and now computers, determine the rate at which the amplitude of the P- and S-wave peaks die off (get smaller) with distance from the epicenter, according to a model based on many recordings of many earthquakes.

One useful thing about using magnitude as a measure of earthquake size is that a simple number between 0 and 10 can be used to describe events within the entire practical size range of earthquakes, from the smallest commonly recorded quakes to the largest, so far. Magnitude, as opposed to intensity, has become the scale of choice for expressing earthquake size. However, when we measure an earthquake's magnitude directly, based on the recorded amplitude of seismic waves, we must average the readings taken at different locations. This takes time and sometimes leads to confusion, particularly in the media. For example, the geology beneath a seismic station that houses a Wood-Anderson seismometer can affect the amplitude of the recording. An instrument installed on soft sediments will invariably produce larger amplitudes, as compared to one installed on solid rock. An earthquake will also often radiate more energy in some directions than in others. Although they may differ slightly, estimates between stations usually agree within about a half a unit, or 0.5. Richter and Gutenberg originally quoted magnitudes only to the nearest

0.5, based on the precision of their methods, and because they only operated a few stations. Nowadays, the press and public are greedy for that decimal place, so seismologists average the readings produced by many stations. The average is generally a pretty reliable number. A few stations can be dropped or added, and the magnitude will stay about the same. Sometimes, however, a preliminary magnitude is issued before all the data have come in and eventually must be adjusted.

Over the years, earthquake observatories and labs have supplemented, and in some cases replaced, the original local magnitude scale with several other methods of determining magnitude. Local magnitude, as its name implies, only applies to local earthquakes, or those that occur within about 450 kilometers (280 miles) of a station. Also, practically speaking, the local magnitude scale only works reliably for earthquakes with magnitudes between 2.5 and 6.0. Below 2.5, few Wood-Anderson seismometers can detect and record a quake. And above about 6.0, although earthquakes produce larger amplitudes, the increases recorded on the Wood-Anderson instrument are deceptively small. These larger earthquakes produce many very long-period waves, characterized by a slow rocking motion rather than the sharp jolting that we normally associate with an earthquake. These are waves that Wood-Anderson seismometers cannot record. As an alternative, well-recorded surface waves from shallow earthquakes are sometimes used to calculate a surface wave magnitude. And in deep earthquakes, which do not generate large surface waves, the amplitude of the P-waves or S-waves can be used to calculate the body-wave magnitude. These types of measurements are particularly useful when determining the size or magnitude of distant quakes far beyond the range of local magnitude.

For the smallest of earthquakes, other methods come into use as well. A duration magnitude can be computed from the total duration of the seismic disturbance on the seismographic record, or the coda magnitude can be determined by the time it takes for the S-wave activity to return to its pre-earthquake or background level.

So each quake can have two or three different magnitudes: local, body, and surface wave. Which is right? They all are. Each is just a different way of describing an earthquake's size or its "strength". In reality, none of these magnitudes is an actual physical measure of an earthquake's energy or power.

The search for a physically meaningful measure of earthquake size led seismologists to develop yet another type of magnitude, the moment magnitude. Moment magnitude is not arbitrarily based on amplitude readings like the others, but on a physical quantity known as the seismic moment. The seismic moment is simply a measure of how much and how far the Earth's crust shifts during a quake, and is related to the amount of strain energy released. Moment magnitude is proportional to the logarithm of the seismic moment, meaning that the amount of strain energy released is mathematically transformed into a format similar to the other types of magnitudes. Moment magnitude works for earthquakes of any size or depth. The one problem with using moment magnitude is that it takes a bit more time to determine than the other types. An earthquake's seismograms must be matched, wiggle for wiggle, with a "model," or computerized description, of the quake.

At Caltech's Seismological Laboratory, moment magnitude is used, at a minimum, for all earthquakes above 6.0 (in moment magnitude). Since the determination of moment magnitude takes quite a bit of computation, earthquakes with magnitudes below 6.0 are still generally reported using the Richter and Gutenberg scale of local magnitude. However, all the Wood-Anderson seismometers have been retired and replaced with modern, "broadband" digital seismometers. Ironically, to report true local magnitudes, the new seismometer's recordings must be converted by a computer into what it would have been using one of the old Wood-Anderson instruments (remember this is an inherent part of the definition of Richter's local magnitude).

The news media, and hence the general public, have somehow lumped all the magnitude scales together under a general, now famous term: the Richter scale. The magnitude of choice is usually the largest one available, which-fortunately, in the case of the largest earthquakes-is the accurate moment magnitude. One almost never hears a seismologist use the phrase "on the Richter scale." They use the term *magnitude* instead, usually prefaced with a descriptor indicating how the measurement was made: moment magnitude, surface wave magnitude, etc.

The World's Largest and Smallest Quakes

Since the invention of the seismograph more than a century ago, the largest earthquake ever observed had a moment magnitude of 9.5. This quake to beat all quakes occurred in 1960 in Chile, within the subduction zone beneath the Andes

Mountains on the west coast of South America. Chile's colossal quake also triggered a devastating tsunami that crashed onto the shores of Hawaii, causing major damage to the town of Hilo. The century's second largest earthquake was the 1964 Good Friday quake in Alaska. This massive trembler had a moment magnitude of 9.2 and caused extensive damage in Anchorage and throughout the region. Once again, Hawaiians suffered from the onslaught of gigantic waves as the 1964 Alaskan quake generated another tsunami. A surface wave magnitude of 8.5 is often quoted for the Good Friday quake, illustrating the confusion caused by having multiple magnitude scales. Unless otherwise specified, throughout the remainder of the text, moment magnitude will be used in reference to quakes larger than 6.0.

To produce an earthquake with a magnitude of 9 or greater, a fault's rupture must extend over a very large surface area. Generally, such massive quakes indicate that the fault rupture extends for more than 1,000 kilometers and has occurred in a subduction zone where the subducting slab descends at a very shallow angle. It takes four to five minutes for 1,000 kilometers of fault to rupture. Thus these megaquakes not only produce extremely intense ground shaking, but the shaking lasts for a relatively long time. The disastrous Alaskan quake reportedly lasted a very long four minutes.

California's San Andreas fault is considered incapable of such a large earthquake, fortunately. Nonetheless, damaging earthquakes up to a magnitude of 8.0, approximately the size of the 1906 San Francisco and 1857 Fort Tejon earthquakes, are possible. They have happened in the past, and since plate motion continues unabated, are inevitable in the future. A particularly likely spot for the next big earthquake in North America is within the Cascadia subduction zone, which lies between the northern end of the San Andreas fault at Cape Mendocino, California, and Vancouver Island, British Columbia. (refer to *Figure 2.6*) Here, a subducting slab descends ominously at a shallow angle. Major cities like Vancouver, Seattle, and Portland sit precariously atop the Cascadia subduction zone. There have been no monster earthquakes here in the recent past, but recorded history is relatively short. The geology of the region suggests that it could happen, and indeed did happen long ago. Japanese records tell of a monster tsunami on January 27, 1700. Based on the characteristics of this tsunami, it could have been triggered by an earthquake within the Cascadia subduction zone. Geologic studies further suggest that the average recurrence time between monster quakes in the Cascadia subduction zone is on the order of about once every 300 to 400 years-so Seattle is probably not due just yet. However, earthquakes do not recur like clockwork, and it is always best to start preparations sooner rather than later.

Small earthquakes are much more common than large ones. In fact, the more numerous and sensitive the seismic stations, the more numerous and smaller the earthquakes that are recorded, particularly if the seismometers are installed in remote areas to reduce outside noise, such as traffic and wind. The Southern California Seismic Network, one of the most sensitive regional networks in the world, can, in some areas, record most quakes above magnitudes of 1.2 to 1.5. Even in a so-called quiet year, almost 10,000 earthquakes are recorded annually within the California network. It is possible for people to feel quakes as small as 1.8 to 2.0, but the motion is so slight that such quakes usually go unrecognized. An earthquake generally needs to be about a 2.5 to 3.0 in magnitude to get the public's attention. In California, a 4.0 quake will normally trigger anxious phone calls to Caltech's Seismology Lab, and a 5.0 will break windows and shake products off the shelves in grocery stores.

Our ability to measure and record more and smaller earthquakes can lead to some fascinating, although potentially confusing, statistics about the Earth's seismicity. In the 1930s, when Richter and Gutenberg were studying earthquake magnitudes, the Southern California Seismic Network had only seven recording stations. Because it takes recordings from at least three seismic stations to locate a quake, most of the earthquakes listed in the catalog were 3.0 or larger. In areas of the network where stations were few and far between, quakes had to be even more significant, up to about 4.0. The present Southern California network contains approximately 300 stations, and earthquakes with a magnitude of 1.5 are routinely recorded and located (remember that a 1.5 quake is 100 times smaller than a 3.5). The result is that the region's present-day catalog of earthquakes contains many more seismic events than in the past. Does this mean that the Earth's seismicity has increased? No, the apparent change observed between the 1930s and the present reflects the increasing sensitivity of our seismic networks.

Examination of California's rather short earthquake history does, however, reveal some spectacular variations in the number of quakes that occur annually, mostly because of the large number of aftershocks following particular events. Are these changes leading up to the Big One? Probably not; over the long term, the seismicity rate in California appears to have stayed about the same. But to adequately judge whether or not the seismicity rate has changed, we need to choose a specific length of time and compare a fairly complete record of earthquakes within a specified range of magnitudes. And it should be kept in mind that even 100 years of seismographic history is little more than a moment in time, geologically speaking. Deciphering earthquake history is like having several pieces of a jigsaw puzzle and trying to construct the puzzle's picture based on just those few pieces.

Exactly how many small quakes are there, relative to the number of large ones? It varies from region to region, but fortunately there are always more small than large quakes. A general rule of thumb is that with each whole step down in magnitude, there are ten times more quakes. In other words, for every quake of a magnitude 3.0, there are ten magnitude 2.0 earthquakes, and for every magnitude 4, ten magnitude 3.0 earthquakes, and so on. Seismologists love logarithms, so they have defined a number called the b-value to quantify the ratio of small to large quakes. A b-value of 1.0 means, mathematically, that there are ten more quakes for each decrease of one unit in magnitude. The b-value tends to vary, with location and possibly with time, between about 0.8 and 1.2.

Knowing the seismicity rate (how many earthquakes per year) and the b-value allows us to extrapolate into the future and estimate how many earthquakes of a specified magnitude a region will experience in a given length of time. This information is critical to structural engineers when they are designing large buildings, bridges, or dams, particularly in earthquake-prone areas. Inevitably, the total number of earthquakes above any magnitude varies significantly from year to year.

Mainshocks, Foreshocks, and Aftershocks

Earthquakes very seldom occur alone. Most earthquakes are followed by numerous aftershocks, and some are preceded by foreshocks. By definition, the largest quake in a sequence is the mainshock, occurring between any foreshocks and aftershocks. For example, in the Northridge earthquake the mainshock, the largest, had a moment magnitude of 6.7. There were no foreshocks, but immediately after the mainshock and continuing for about five years there were more than 14,000 aftershocks. Thirty-six percent of the aftershocks occurred in the first month, which is typical. Aftershocks usually have an orderly and steady rate of decay which means that they become less frequent with time. This does not mean that aftershocks necessarily decrease in magnitude with time. The ratio of large to small aftershocks is similar to that for all quakes (expressed by the b-value). It is possible to have a large and damaging aftershock several months after the mainshock, but usually any larger aftershocks come sooner rather than later, just because all aftershocks are more likely to occur sooner rather than later.

How do we know when the aftershocks are over? We don't, but we can make a pretty good educated guess based on a region's average seismicity rate prior to the beginning of the sequence (the mainshock or the first foreshock). The aftershock sequence is considered over when the rate of quakes decreases to the background (or prior) level. Sometimes this practice causes a little consternation on the part of the public and the media. As luck would have it, there is often a notable aftershock later than many expect. Again, it is not as likely to happen as in the first week, but later and large aftershocks do occur.

The epicenters of the immediate aftershocks usually define a clear zone of activity, called the aftershock zone. Most of the time, the aftershock zone outlines the section of the fault that has ruptured during the mainshock. A fault rupture can begin at one end and progress along the fault (a unilateral rupture), or the tear can start in one place and travel in two opposite directions (a bilateral rupture). Sometimes the mainshock also triggers activity on other, adjacent faults. If this occurs within a few hours or days, the resulting earthquakes are counted among the first quake's aftershocks. In the Northridge quake, almost all of the aftershocks happened on the same fault as the first rupture. Here the fault plane dipped down at an angle of about 45 degrees to the Earth's surface. The mainshock occurred at the southern and deepest end of the fault, meaning that in the quake's fifteen or so seconds of shaking, the rupture started at the bottom and moved upward through the earth and to the north. The aftershocks clearly defined the rupture plane until it got to a depth of about 5 kilometers beneath the surface. Then the rupture began to branch out into a number of smaller cracks, making the northernmost aftershocks define more of a cloud than a plane.

Some of the locals within the Santa Monica Bay area may refute the earlier statement that there were no foreshocks in the Northridge quake. In Santa Monica Bay, during the week before the mainshock, there were several small but felt earthquakes. Seismologists are still arguing about whether those particular quakes were physically related to the Northridge earthquake or if their occurrence was just coincidental. In either case, they were, by definition, not foreshocks, as they did not occur in the same place as the mainshock, or even within what later became the aftershock zone.

A more complicated example of foreshocks, mainshocks, and aftershocks is the other big and relatively recent earthquake sequence in southern California. The Landers quake occurred on June 28, 1992, had a moment magnitude of 7.3, and was the largest earthquake in southern California in the last forty years. However, because the Landers quake occurred in a sparsely populated area of the High Desert, it received little press attention, unlike the Northridge quake. The small towns of Landers, Yucca Valley, and Joshua Tree and the mountain resort areas of Big Bear and Lake Arrowhead were the most severely effected. The mainshock killed only one person and did much less property damage than the Northridge quake. Also, at the time of the quake, the news and the attention of the public were focused on other pressing problems in Los Angeles, civil unrest and rioting.

The Landers story really began with an earthquake about two months earlier, in April 1992, in the Joshua Tree National Monument. It was a textbook magnitude 6.1 earthquake, with a few foreshocks and lots of aftershocks. The fault rupture was about 15 kilometers long, but remained hidden within the Earth. The aftershocks died off with time as expected, but at a rate slower than normal for a southern California sequence. Aftershocks can have their own aftershocks, and often there are little bursts of activity throughout the entire aftershock sequence. One of these little bursts occurred during the night of June 27, 1992. And on the following morning, at the same location, another quake struck, but this time it was much larger, with a moment magnitude of 7.3. That earthquake was named the Landers quake, and the Joshua Tree quake became a foreshock to the Landers. The Joshua Tree quake's foreshocks became foreshocks to a foreshock, and Joshua Tree aftershocks became Landers's foreshocks. Eventually, the entire sequence of earth-shaking events became known as the Joshua Tree/Landers sequence. But then the Landers sequence became even more complicated. The largest aftershock of Landers occurred off the main rupture, in the nearby San Bernardino Mountains under Big Bear. This event occurred about two hours after the Landers quake, had a moment magnitude of 6.2, and did most of the property damage for the whole sequence. With this later event, the entire episode became the Joshua Tree/Landers/Big Bear sequence, or simply the Landers sequence.

Of course the Big Bear quake added its own aftershocks to the sequence, and now, some seven years after it started, the grand total of associated quakes amounts to about 65,000. The Landers sequence is a rare and scientifically valuable data set. Seldom does a magnitude 7+ earthquake occur smack in the middle of a high-quality seismic network, and one that has digital archiving available through the Internet. Five years after the Landers sequence, network seismologists had accurately located and assigned a magnitude to all the related tremors. When plotted on a map, the Landers sequence appears like a Greek letter lambda (λ) on the southern California desert (*Figure 2.8*). The longest "arm" of the lambda outlines the Landers rupture (with the Joshua Tree rupture hidden inside

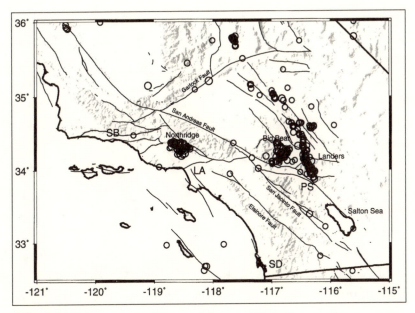

Figure 2.8
Map of southern California showing major faults and locations for earthquakes magnitude 4.0 or larger, 1992 through 1998. LA, PS, SB, and SD refer to Los Angeles, Palm Springs, Santa Barbara, and San Diego, respectively.

the southern end of it) and the shorter "arm" outlines the Big Bear rupture.

Now consider the role of aftershocks, foreshocks, and mainshocks in determining a region's seismicity. Typically, a "declustered" catalog of earthquake events is used to compute a region's seismicity, meaning that for each foreshock/main-shock/aftershock sequence, only the mainshock is counted. From the Northridge and Landers quakes together, that would mean ignoring 79,000 smaller events!

Aftershocks are a continuation of the process known as elastic rebound. Imagine a piece of paper crumpled into a tight ball and then set on a table. Left to its own devices, the ball of paper slowly uncrumples and enlarges slightly in size. For awhile the wad continues to pop and crackle, expanding a bit more in the process, until eventually it reaches a stable configuration. This is analogous to the aftershock sequence, with the Earth's crust crackling and shifting after an earthquake until it finally reaches a stable situation.

Even the smallest of earthquakes releases strain within the Earth's crust. The release of tension on one portion of a fault can rearrange the strain along other portions of the same fault or on separate, nearby faults. Strain usually increases at the edges of a quake zone; in front of the rupture strain has not been released, while behind it, it has been. Consequently, the tip of any recent rupture becomes a good candidate for seismic activity in the future. This process is similar to the way a crack in a car's windshield grows longer. Any place along a fault rupture is a candidate for strain concentration, especially if the fault plane is not perfectly smooth (which it never is). An earthquake may also significantly affect nearby faults, their strain either increasing or decreasing depending on their location and orientation relative to the mainshock rupture. If the tension on a fault exceeds its strength, then an aftershock occurs. A large earthquake essentially creates a large-scale disruption of the surrounding strain field and can thereby trigger numerous aftershocks. In the case of the Landers quake, the passage of seismic waves triggered small quakes 440 kilometers away at Mammoth Lakes. Large-scale strain changes usually occur up to about one fault rupture length away.

Swarms and Clusters

In some locations, particularly within volcanically and geo-
thermally active areas, the usual sequence is so complex that it
is called an earthquake "swarm" rather than a mainshock/after-
shock sequence. In regions of geothermal activity, molten rock
below the Earth's surface heats the local groundwater, and
often at the surface, creates natural hot springs or steaming
geysers. In California, geothermal activity occurs in Imperial
Valley, the Coso Ranges near Ridgecrest, and the well-known
ski resort of Mammoth Lakes. A typical Imperial Valley earthquake swarm starts with a few sporadic, small seismic distur-
bances over a few days' time. The seismicity rate slowly
increases until, within the span of just a few days, there are
hundreds or thousands of small quakes, some of which may
reach a magnitude of about 5.0. An earthquake swarm does not
begin with a mainshock, or even a few foreshocks followed by
a mainshock. There are often three or four candidates for the
title of mainshock, timed in the middle of the swarm. The seismicity then gradually decays like a regular aftershock
sequence. Another term used now and then is *cluster*. A cluster
of earthquakes is usually small, with up to one to two dozen
members. If the largest quake is not the first, the entire
sequence may be referred to as a cluster.

Predicting Aftershocks

It is difficult to tell when aftershocks are really aftershocks and
when they are really foreshocks leading up to a new mainshock. If a magnitude 5.0 quake occurs, how do we know that it

is not really a harbinger of a later, larger quake? Unfortunately, a seismogram can't help us here. The relationship in time and space between other quakes can only be determined after the entire sequence is over. But a few hours into the earthquake activity, seismologists can use the magnitude of the mainshock, the rate of quakes, the ratio of small to large quakes, the rate of decay, and a generic sequence, specific to the region, to quickly estimate how many and what size aftershocks to expect. Unfortunately, statistics cannot be used to pinpoint exactly when or where damaging aftershocks will occur. In the past few years, following any major quake, it has become common practice for the U.S. Geological Survey to issue an aftershock alert that describes in general terms how many aftershocks, and of what size, can be expected.

What are the chances that a mainshock will have an aftershock bigger than itself, and thereby become a foreshock (like the Joshua Tree quake)? In the generic southern California sequence, this happens about 5 percent of the time-the odds are one out of twenty.

Another interesting sequence of California tremblers was the Superstition Hills event in November 1987. It started with a magnitude 6.2 quake around 5:00 in the evening on November 23, in the middle of the Imperial Valley. The fault rupture extended diagonally across the valley, from southwest to northeast, just south of the Salton Sea (*color plate 8*). Near the northeastern end of the rupture is a tiny hamlet, called Bombay Beach. Bombay Beach is at the southern end of the Coachella Valley segment of the San Andreas fault, a section that last ruptured around 1680. Seismologists worried that the strain in this segment was very high and that if the initial 6.2 quake increased it sufficiently, a major rupture along the main San Andreas fault could occur. At Caltech's Seismology Laboratory, scientists and staff remained nervously alert overnight, and early the next morning their predictions came true, to an extent-a larger magnitude 6.6 quake struck. But luckily, it happened at the other end of the fault rupture, in a deserted area called the Superstition Hills. Even though this is one of the seismology lab's more successful tales, it illustrates the sorry state of earthquake prediction. Imagine the panic if the staff had sent out an alert to everyone in southern California predicting an imminent earthquake.

The Breaking Up of Faults: Segmentation

The San Andreas fault can be divided into several individual segments that may or may not participate in any particular earthquake, depending on how much strain has accumulated along each at the time. Fault ruptures, or large earthquakes, tend to begin and end at segment boundaries, although they do not always do so. The Coachella segment extends from Bombay Beach, on the shore of the Salton Sea, to the San Gorgonio Pass west of Palm Springs. This region of the San Andreas is currently thought to be under the most strain. Geologic evidence suggests that the last rupture in this area occurred around 1680. Because of the long time since its last big quake, common sense suggests that this part of the San Andreas will be the next to go. The date of rupture is anyone's guess, but areas at risk, like the towns of Indio, Palm Springs, and perhaps San Bernardino, might do well to begin reviewing their preparedness plans.

The Seismic Cycle

Within the Earth's crust there is an eternal cycle through which the plates move, strain accumulates along faults, and rupture occurs releasing the strain. Compared to the human lifetime, or even the timespan of some cultures, the seismic cycle is long. Only in China and Japan, where written records of earthquake damage exist from three to four thousand years ago, can some sense can be made of long-term seismicity patterns. In California, the written record is barely 250 years long, and so what happened previously in terms of seismic activity is unclear.

The last "great" San Andreas earthquake (magnitude of 8.0 or larger) was the 1906 San Francisco quake, which ruptured a section of the San Andreas fault that was about 430 kilometers (268 miles) long, from the Santa Cruz Mountains all the way to Cape Mendocino. Prior to that, there was a great quake in southern California in 1857, the somewhat less famous Fort Tejon earthquake. The Fort Tejon earthquake also ruptured a long section of the fault, some 360 kilometers in length. What happened before 1857? Here historical records become sparse at best. Populations within the area were very thin prior to the 1850s, and written accounts are incomplete. Reports do suggest, however, that there were two large southern California earthquakes in 1812: one damaged the Santa Barbara Mission and the other destroyed parts of the San Juan Capistrano Mission. From damage reports alone, there is no way to guess which fault caused either. However, there are some very old trees growing within the San Andreas fault near Wrightwood, California. Samples cored from the wood of the trees show abnormally narrow growth rings around 1857 and 1812. Scientists interpret this to mean that during 1857 and 1812 something unusual happened that caused large sections of these trees to break off. Wood samples from trees located slightly off the fault zone show that they were unaffected. From this evidence, we can deduce that the 1812 San Juan Capistrano quake was probably on the San Andreas.

When written records are absent, seismologists must look for clues elsewhere to deduce ancient earthquake events. As Paleoseismologists, they study trees, sediments, and old stream or lake beds that lie along a fault. One particularly famous lake (sometimes lake, sometimes swamp, and sometimes dry) deposit is located along the San Andreas fault at Pallet Creek near Palmdale, California. Trenches dug within the lake's layered sediments have revealed a repeated pattern of disruption in the bedding. From this, scientists have identified a dozen or so prehistoric earthquakes, estimated their offsets, and used carbon-14 dating of bits of vegetation in the sediment beds to estimate the dates they occurred.

In principle, the theory of elastic rebound suggests that the same section of every fault should rupture at very regular intervals, since the plates move at a relatively steady rate, and each fault presumably has a particular breaking strength. Unfortunately, Mother Nature does not seem to operate as simply as theory predicts. Tree ring data suggest that in Wrightwood, quakes occurred in 1812 and 1857, only forty-five years apart. But it has now been 143 years since a major quake last rocked the area. In fact, the paleoseismic record for nearby Pallet Creek suggests that there have been quakeless intervals lasting some 250 years. What would appear to be a useful tool in anticipating future disasters, namely the seismic cycle and elastic rebound theory, turns out to be of limited use. The problem appears to lie in the complexity of the Earth's geology.

California harbors dozens of major faults (*color plate 8*), but is also fraught with an abundance of smaller faults. Each earthquake rearranges strain on the faults-relieving strain on one and increasing it on another. The result is a chaotic history of unpredictable earthquakes, rather than a nice smooth seismic cycle that operates like clockwork. Consequently, seismologists must rely on statistical probabilities when making statements about the future. For example, the chances of a magnitude 7.5 or larger quake (the "Big One") on the San Andreas is at least 50/50 in the next thirty years. An individual deciding whether or not to move to California (or out of it!) might not find this information very helpful, but an engineer designing a bridge, large building, or anything expected to last at least thirty years must take the possibility of strong ground-shaking very seriously.

Earthquake Prediction: Is It Possible?

For the most part, seismic statistics leave people, including emergency services personnel, somewhat unsatisfied. When it comes to earthquakes, they want to know exactly where, when, and how big. In the daily weather report, we would rather hear "It will rain" or "It will not rain" than a "There's a 50 percent chance of rain." What does a 50 percent chance of rain really tell us? It is just as likely to rain as not.

Can seismology ever produce predictions as accurate as the weather forecasts? Probably not. The success of weather predictions is partly due to the availability of a huge number of observations of the Earth's atmosphere, at the surface, at various altitudes, and from orbiting satellites. It also helps that, except for occasional clouds, the atmosphere is transparent and we can see what is going on. Also, the processes involved in the weather are fairly well understood. Seismologists are not so lucky. They cannot see down deep inside the Earth and witness what is happening from hour to hour or day to day. Without time-consuming and expensive drilling operations, they can only make measurements at the Earth's surface and infer what is happening below from seismic waves or measurements of deformation. Furthermore, too many parts of the geologic puzzle remain poorly understood; for instance, we are still unsure as to how earthquakes start and what limits their size.

Some seismologists believe that most earthquakes will never be predicted, because earthquakes are inherently unpredictable! Imagine trying to break a board over your knee, and suppose that you are trying to predict exactly where and when it will snap. The force that you exert on the ends of the board is something like the stresses induced by the motion of the Earth's plates. The bending of the board is analogous to the elastic strain. Initially, some small cracks in the board may start to form (foreshocks), and these might help you predict where the final break (mainshock) will happen, but they will not be of much help to you in determining exactly when it will occur. The final rupture will probably start as one of those little cracks, indistinguishable from any of the others, and occur only when the strain is high enough that the small crack spreads across the entire board. The only thing you can really do is be prepared: don't put your face close to the board while applying the stress.

While scientists cannot say exactly where and when an earthquake will strike, they can now gather enough evidence to determine the probability of a quake of a particular magnitude occurring within a specified time frame. For instance, in Southern California, scientists estimate that the probability of a magnitude 7 or greater earthquake by the year 2024 is as high as 80 to 90 percent.

Quake Precursors: Truth or Myth?

In the 1970s, there was great optimism about earthquake prediction. A few so-called earthquake precursors had come to light, and there was even a theory (known as dilatancy) put forth to explain many of the phenomena that come before a large earthquake. A series of foreshocks is an example of a precursor. However, since foreshocks look just like any other earthquakes, they are not in themselves very useful in prediction. From all points around the globe, there are numerous anecdotal reports about other precursors, earthquake folklore, if you will.

Many widely reported earthquake precursors are related to groundwater. A few hours before a large earthquake, marked changes have been reported in the level or flow of wells and springs. Groundwater has also reportedly changed temperature, become cloudy, or acquired a bad taste. Occasionally, electrostatic phenomena such as earthquake lights (similar to the St. Elmo's fire that appears on ships during electrical storms), and changes in the local magnetic field have been reported. Anecdotal reports also persistently include the strange behavior of animals, which might be linked to

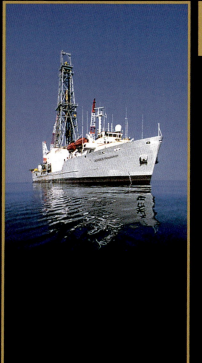

Color plate 1
Ocean drilling ship
Courtesy Ocean Drilling Program.

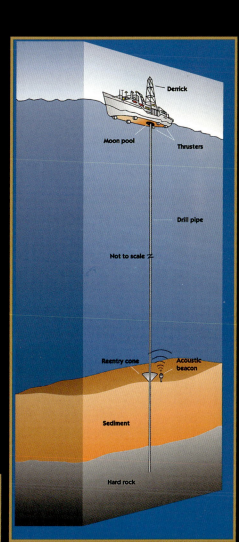

Derrick

Moon pool

Thrusters

Drill pipe

Not to scale

Reentry cone

Acoustic beacon

Sediment

Hard rock

Color plate 2
Sketch of drill ship operations
Courtesy Ocean Drilling Program.

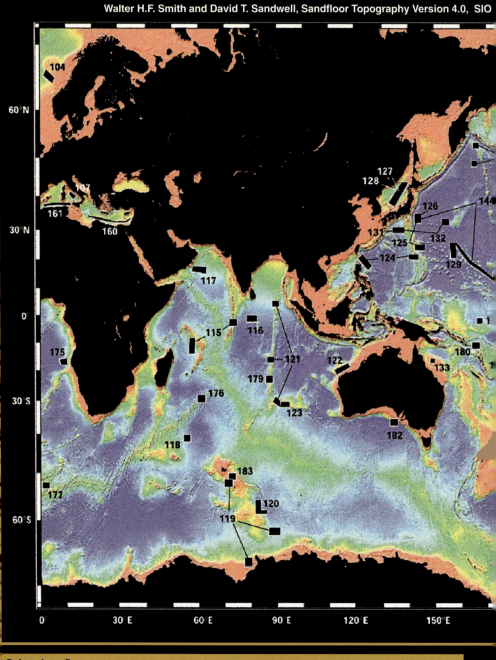

Color plate 3

Seafloor topography (courtesy Walter H.F. Smith and David T. Sandwell) with an overlay of ocean drilling locations, 1985–1998 (courtesy Ocean Drilling Program). Range of depth: Orange is shallowest, with depth increasing to yellow, green, and blue.

September 26, 1996   Copyright: 1996, Walter H.F. Smith and David T. Sandwell.

World Seismicity: 1975–1995

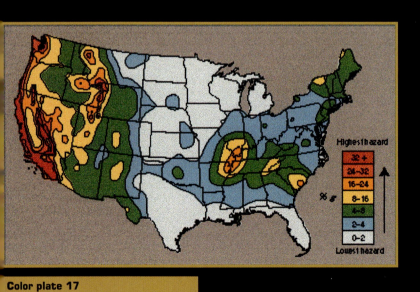

DEPTH

Color plate 4
Worldwide seismicity, 1975-1995. Colored points reflect the depth of the earthquake.
Courtesy U.S. Geological Survey.

Highest hazard

32+
24–32
16–24
8–16
4–8
2–4
0–2

% g

Lowest hazard

Color plate 17
Seismic hazards map shows a 10 percent probability of experiencing severe ground-shaking over a 50-year period, as a percentage of g (gravitational acceleration).
Courtesy U.S. Geological Survey.

(5a)

(5b)

Color plate 5

(a) The submersible Alvin (courtesy OAR/National Undersea Research Program, NOAA). (b) Black smoker at a deep-sea vent (courtesy OAR/National Undersea Research Program). (c) Vent tubeworms (© Woods Hole Oceanographic Institution).

(5c)

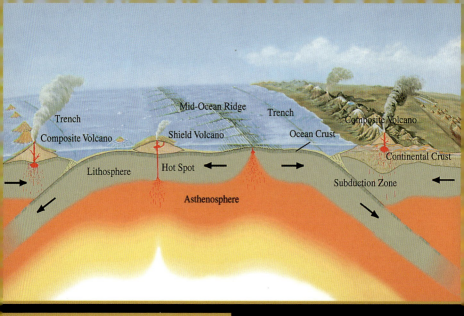

Color plate 6
Diagrammatic sketch of plate boundaries
and tectonic processes.
*After Vigil, J.F., Wall map — This Dynamic Planet, US
Geological Survey, Smithsonian Institution, and U.S. Naval
Research Lab.*

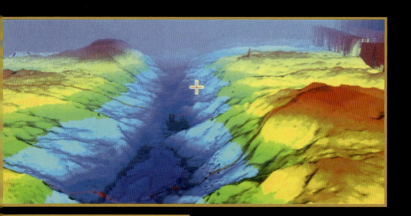

Color plate 22
Computer generated image of the
Mid-Atlantic Ridge (©Woods Hole
Oceanographic Institution).

Color plate 18
Damage to an apartment building with a "soft first story" during the Northridge Earthquake.

Color plate 7
San Andreas fault, California.
Courtesy U.S. Geological Survey.

Color plate 23
Damage to unreinforced masonry building during the Northridge Earthquake.

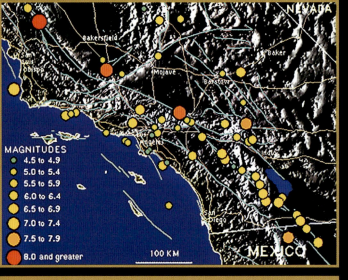

Color plate 8
Map of Southern California showing faults (light blue) and highways (tan), and an overview of relatively recent earthquakes. Moment magnitudes are used for quakes greater than 6.0, and local magnitude for those below 6.0.
Courtesy U.S. Geological Survey.

MAGNITUDES
- 4.5 to 4.9
- 5.0 to 5.4
- 5.5 to 5.9
- 6.0 to 6.4
- 6.5 to 6.9
- 7.0 to 7.4
- 7.5 to 7.9
- 8.0 and greater

100 KM

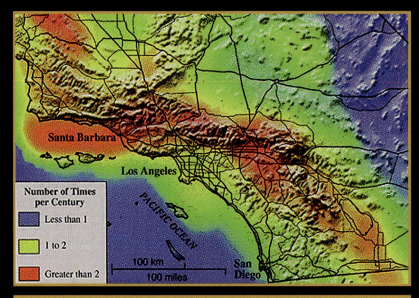

Color plate 9
Probability of earthquakes in Southern California that will produce severe ground-shaking.
Courtesy U.S. Geological Survey.

Number of Times
per Century
- Less than 1
- 1 to 2
- Greater than 2

100 km
100 miles

(10a)

(10b)

Color plate 10
(a) Hawaiian lava flow and (b) fountain surrounded by black basalt.
Courtesy U.S. Geological Survey.

(11a)

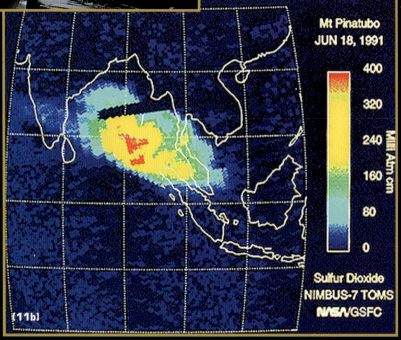

Mt Pinatubo
JUN 18, 1991

400

320

240

Mill Atm cm

160

80

0

Sulfur Dioxide
NIMBUS-7 TOMS
NASA/GSFC

(11b)

Color plate 11
(a) June 1991 eruption of
Mount Pinatubo and (b) the
resulting satellite-derived
sulfur dioxide concentration in
the overlying atmosphere.
Courtesy NASA, U.S. Geological Survey.

Color plate 12
(a) False-color radar image and (b) map of Mount Pinatubo. On the radar image, red denotes ash debris near the summit, and dark colors outline mud flows radiating off the volcano. *Courtesy NASA, U.S. Geological Survey.*

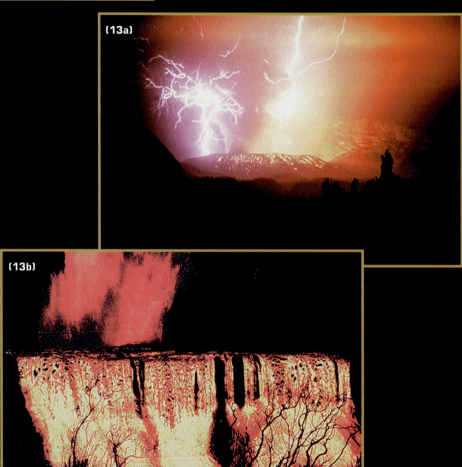

Color plate 13
(a) Nighttime lightning over an erupting volcano and (b) lava falls, Kilauea, Hawaii. *Courtesy U.S. Geological Survey.*

(13a)

(13b)

(14a)

(14b)

Color plate 14
(a) Lake Nyos, Cameroon, and
(b) aftermath of lethal gas
release.
Courtesy U.S. Geological Survey.

(15a)

(15b)

(16a)

(16b)

Color plate 16
After the 1998 Papua New Guinea tsunami.
(a) Sand spit fronting hard-hit Sissano
Lagoon; (b) destroyed home; (c) bent wood
supports and bucket high in a tree;
(d) computer simulation of the tsunami.
(courtesy U.S. Geological Survey).

(16c)

(16d)

Color plate 19 Leading depression wave of tsunami striking Manzanillo in 1995.

Color plate 20 Debris from a coastal village in Babi piled high in a coconut grove.

Color plate 21 Damage from 1994 tsunami in East Java.

electrostatic phenomena or small foreshocks. Changes in strain and creep (silent motion, without an accompanying earth-quake) along a fault normally locked by friction could also be considered precursors.

In China in the 1970s, it became popular for people to pre-dict earthquakes using "backyard" measurements, such as the monitoring of well levels and observation of farm animals. At least one earthquake, the Haicheng quake in 1975, was suc-cessfully predicted and a town evacuated, proving that, at least in some cases, earthquake prediction is possible. The Haicheng earthquake had hundreds of foreshocks, making it an easier-than-average quake to predict. Groundwater changes and anomalous animal behavior were also reported (for example, hibernating snakes supposedly awoke and froze to death). In China, "evacuation" meant that compulsory outdoor movies were shown, so that when the quake did happen and the town was severely damaged, no one was killed. But Chinese seismol-ogists missed predicting the catastrophic Tangshan earthquake in which at least 250,000 people reportedly perished.

Physical measurements, rather than backyard observa-tions, are more heavily relied upon within the United States. In the 1980s, the town of Parkfield, California, was singled out as a natural laboratory in which to study earthquake precursors. Parkfield is in central California, right on the San Andreas fault. To the northwest of town, the fault seems to creep more or less continuously and has never accumula-ted enough strain to produce any large earthquakes. However, to the southwest, Parkfield borders on the so-called Carrizo Plain segment of the fault, right where the 1857 earthquake rupture probably began. This segment appears to play host to a magnitude 6-ish earthquake, on average, about every twenty-three years. In the 1980s, the twenty-three-year cycle was coming due, so Parkfield seemed like the ideal place to install instruments to observe earthquake precursors. Parkfield literally became an outdoor earthquake laboratory and was equipped with all manner of seismometers, creepmeters, strainmeters, tiltmeters, etc. But it is now 1999, and the "Parkfield earthquake" has not yet happened. Nor have any obvious precursors been observed, although there has been some creep and a few small quakes.

Unfortunately, many of the possible precursors are, for one reason or another, problematic. So far, we know of no way to distinguish a foreshock from any other earthquake. Well levels and spring flows are affected sufficiently by rainfall to obscure any possible precursory changes having to do with movements of the Earth's crust. In developed regions full of power lines and radio interference, electrical and magnetic changes are almost impossible to measure accurately. And strange, precursory animal behavior is hard to observe and distinguish in animals that are affected by so many different stimuli.

Even if a reliable earthquake prediction were to be made, what the subsequent response should be is, so to speak, on shaky ground. Unquestionably, many lives could be saved if emergency services personnel were all on duty, various hazardous material manufacturing operations were halted, and so forth. But there is great concern that the warning of an impending quake would cause immense panic and result in the traffic jam to end all traffic jams (the short earthquake history in California has clearly taught us that the freeway is one of the worst places to be caught in an earthquake!), not to mention the possibility of rioting, looting, and general anarchy.

Should an earthquake prediction be made, particularly for California, there is already an established response protocol. Two different groups come into play: the California Earthquake Prediction Evaluation Council (CEPEC) and the National Earthquake Prediction Evaluation Council (NEPEC). Each is a panel of highly respected scientists, available on an emergency basis, to review and assess the validity of any earthquake prediction. Expert panels are necessary, because in the past, several less-than-scientific earthquake predictions have grabbed the public's attention. One famous example was that of Iben Browning, who predicted a major quake for New Madrid, Missouri, on December 3, 1990. Browning's prediction was highly suspect, but until scientists stepped forward and denounced it, it threatened to cause a general panic. Still, a certain contingent believed the tabloids over the experts and prepared for the worst.

A number of other, more scientifically based predictions have been presented and discussed before the NEPEC and CEPEC panels. None has ever been deemed sound enough to warrant a public warning. In fact, none of these specific predictions have ever been successful. In Parkfield, specific criteria have been set up in advance by the U.S. Geological Survey (which monitors the Parkfield area) for different levels of warning. For example, if there is an earthquake above 3.2 in the Middle Mountain area, considered to be the likely epicenter of the overdue Parkfield quake, or one above 3.3 in a somewhat larger area, a *D level* alert is issued. Quakes larger than 3.9 and 4.2, respectively, trigger a *C level* alert, and quakes larger than 4.7 and 5.2 trigger a *B level* alert. An *A level* alert requires that the prospective foreshocks match exactly the foreshocks of previous Parkfield earthquakes. By predetermining what has to happen to trigger an alert, the necessity for an emergency CEPEC or NEPEC meeting has been bypassed. The prescribed channel for communications during an earthquake alert in Parkfield begins with the U.S. Geological Survey transmitting to the California Office of Emergency Services, then to the local emergency services people, and finally to the general public. Of course, since Parkfield has only thirty-four residents, the logistics for an earthquake evacuation would be exceedingly simple compared to those for a densely populated city like Los Angeles. What might be of greater concern is the remote possibility of a Parkfield quake triggering a rupture along another segment of the San Andreas, as appears to have happened in 1857.

The bottom line on earthquake prediction is a simple message: don't count on it happening. Research on earthquake prediction is still being conducted at several institutions, but the prospects are not as promising as they seemed after the Haicheng success. Reliable earthquake prediction is clearly complex and remains a thing of the future.

Beyond Prediction: Reducing Risk

Although we cannot reliably predict earthquakes, years of seismologic and geologic research have produced results that are helping to save lives and property during seismic events. Several areas of study have been particularly productive; these include seismic hazard and risk mapping, engineering and the improvement of building design and construction, and an improved ability to detect quakes and rapidly provide accurate information immediately afterward.

Scientists, city planners, engineers, and emergency response personnel are now working together to map seismic hazards and reduce the risks of damage from earthquakes. Seismic hazard mapping begins by studying a region's geology and locating all the faults within the area that could produce earthquakes. A wide array of techniques are used to locate, map, and study fault patterns, including subsurface geophysical studies, seismic recordings, satellite imagery, and paleoseismological research. In the Portland-Vancouver area, in the 1990s scientists also began using aerial geophysical surveys to examine the region's suspected fault patterns. They used a specially designed airplane equipped with instruments to look for distinctive patterns in the Earth's magnetic field, which often reflect underlying faults or tears in the Earth's crust. Survey data confirmed the existence of a fault running under and through Portland, Oregon.

Today, scientists can create overall seismic hazard maps (*color plate 17*), as well as maps that specifically delineate the dangers of ground shaking (*color plate 9*). For a given region, each map shows the expected number of times per century (or other time interval) that the bedrock will shake strongly enough to cause damage. Seismic and shaking hazard maps can be used in numerous ways. They can aid in design requirements for highway bridges, help set insurance rates or environmental standards, or assist in estimating the probability of landslides. Eventually, a three-dimensional image of an area is produced and combined with computer simulations of ground shaking.

Together these help forecast how the ground will shake during an earthquake, and more importantly to define where the shaking will be the greatest. In Seattle, Washington, this information is of particular importance, because the city sits on a basin of sedimentary rock fill that tends to amplify ground shaking.

The type of rock or soil at any particular location has to be factored into an earthquake risk map as well. Soft sediments, and especially uncompacted landfill, greatly amplify ground shaking during an earthquake. Damage tends to be considerably worse in areas underlain by soft sediments or fill, especially where sediments are composed mainly of water-saturated sand. Strong shaking can easily turn these soils literally into quicksand, a process known as soil liquefaction. In the 1989 ("World Series") Loma Prieta earthquake, significant damage occurred in San Francisco's Marina District, 100 kilometers away from the earthquake's epicenter. Ironically, the upscale, trendy Marina District had been built on unstable fill, made largely from the rubble of the 1906 earthquake. In the 1981 Mexico City earthquake, buildings that collapsed in Mexico City, some hundreds of kilometers away from the earthquake's origin, were built on old, dry lake sediments. Structures in the surrounding areas, built on hard rock, incurred significantly less damage.

Seismic hazard and risk maps can also reveal some surprises. In California, the single most active fault, in terms of the largest, most frequent quakes, is, unsurprisingly, the San Andreas. Certainly for those cities and towns located on or near the San Andreas fault, its movement presents the greatest risk. However, the San Andreas fault does not represent a nice, clean break or boundary between tectonic plates, but rather a rough and badly splintered zone of faulting. A map of California's faults looks somewhat like a shattered window. Cracks spread out across the landscape, separating the countryside into many tiny "platelets." It has been estimated that on average, over a long period, about 60 percent of the plate motion in the area occurs across the San Andreas fault. The remaining 40 percent is spread out over all of the other faults in the region. Each of the small faults may have a longer recurrence time for quakes, on the order of several thousand years, compared to a couple of centuries for the San Andreas. However, because there is a huge number of these small cracks throughout the entire region, particularly in places like Los Angeles, Northridge, and Barstow, they may collectively represent a greater risk for damaging quakes.

The Northridge earthquake was a classic example of a non-San Andreas fault quake. In fact, both the 1987 Whittier Narrows quake and the 1994 Northridge event brought to light a whole series of previously unknown blind-thrust faults beneath metropolitan Los Angeles.

Major earthquakes, even if they occur some distance away from a city, can have a devastating impact. For instance, if an 8.0 earthquake were to occur on the San Andreas fault, its impact would undoubtedly be felt some 35 miles away-near greater Los Angeles. The good news is that the short-period ground shaking normally associated with an earthquake would be less intense in Los Angeles. The bad news is that because the rupture could take almost a minute to propagate along the fault, the shaking would probably continue for at least that long. Some buildings that could withstand a few seconds of moderately strong shaking might collapse during a minute of the same punishment. Also, larger earthquakes produce a greater amount of long-period seismic waves than do moderately sized quakes. Long-period seismic waves might not feel like much to an observer at ground level, but they would cause tall buildings to sway back and forth. It is unknown how a minute of long-period shaking would affect downtown L.A.'s collection of towering high-rises.

Earthquakes can also trigger landslides, flooding, widespread fires, volcanic activity, and tsunamis. It is more difficult to predict the risks of indirect or secondary impacts from earthquakes, but all hazards need to be considered when making preparations.

In southern California, a map of the San Andreas fault reveals another major problem. A rupture of a 200-kilometer section of the San Andreas, with some 3 meters of offset, would mean the disruption of all three aqueduct systems (L.A.'s water supply). Additionally, high-pressure gas pipelines, several major freeways, railroads, and high-voltage power lines would all be heavily affected. In emergency services parlance, these vital services are referred to as "lifelines." Even if the amount of physical damage were small, lifeline disruption could cause serious problems.

In addition to improving our ability to assess earthquake risk, seismological research has also helped in the prediction and prevention of earthquake damage. The simplest example is in the difference between how masonry and wood-frame structures withstand earthquakes. Wood-frame houses, a category that includes most houses in the United States, are lightweight and strong. If they are attached to their foundations, most wood-frame houses survive well in an earthquake. Cracks may form in the plaster and chimneys may fall, but on the whole, the damage is usually repairable. Older homes not bolted to their foundations fare less well. However, in quake-prone areas, older structures can be retrofitted for safety.

Unreinforced masonry buildings (called URMs by emergency planners), on the other hand, are very heavy, not very strong, and extremely vulnerable to earthquake damage (*color plate 23*). In California, most URMs are commercial brick buildings. In many other parts of the world, they are traditional stone or adobe housing. In such areas, relatively small quakes quickly turn into major disasters. The Agadir, Morocco, and the recent Colombia earthquakes mentioned earlier are particularly illustrative examples. In California, many of the URMs have been retrofitted, to the extent that the brick exterior walls will not, supposedly, fall dangerously into the streets. Time will tell whether these measures are sufficient. Given that the retrofits are quite expensive and considering the total number of older structures, complete replacement is usually hindered by cost.

In the developed regions of the world, construction is regulated by a set of laws called the building code. In areas unlikely to experience earthquakes, the building code covers general safe construction practices, plus items like fire sprinklers and other safety measures. In areas like California, the building code also includes provisions for resisting earthquake damage. The major risk for buildings during earthquakes is horizontal motion. Most buildings are stronger than necessary in the vertical direction, in case the building is overloaded with contents. But in an earthquake, a building also needs to resist horizontal forces. In California, most new wood-frame buildings have plywood sheathing over the frame, so that in a quake the frame does not deform horizontally. Older homes do not have this modification, and therefore face greater risk in an earthquake.

From past earthquake events, structural engineers have learned the quirks and idiosyncrasies of each type of construction and have updated the building code accordingly. Unfortunately, for economic reasons, the building code is not usually retroactive; it applies only to new buildings. In addition to the building code, there could be, for example, a local or countywide ordinance to retrofit URMs. A little-known fact is that building codes are intended as a *minimum* standard, in the interest of public safety, to preserve life but not necessarily to preserve the building. In the case of a building, such as a hospital, which is expected to function immediately after a quake, more strength and usually added cost are required.

Not surprisingly, earthquakes have a way of pointing out oversights in the building code. The Long Beach earthquake of 1933 started California along the right path. Most of the damage in this quake was to weak and unreinforced structures, which, frighteningly, included many school buildings. Luckily, the quake occurred at around 5 P.M., when the schools were mostly empty. But the quake and its damage to the schools revealed a serious safety problem, so the California legislature passed the Field Act, its first earthquake provision in the building code, which applies to any new public school construction.

A more recent example of problems in the building code is the many so-called soft first story buildings in California. These include a large number of apartment buildings with parking underneath, as well as a number of attractive office buildings that are mostly glass on the first floor. Although the upper floors of such buildings are quite strong, the first floors tend to collapse in an earthquake (*color plate 18*). In the 1971 San Fernando and the 1994 Northridge quakes, the first floors of the Olive View hospital (1971) and the Northridge Meadows apartment building (1994) were both flattened as the floors above, in a single unbroken unit, crashed down. During earthquakes, older, poorly reinforced support columns under freeway bridges have also been known to crumble or fall. This problem became all too clear in both the San Fernando and Northridge earthquakes. An enormous retrofit program was subsequently undertaken to completely enclose bridge support columns in reinforcing steel jackets.

In areas at risk from earthquakes, hard lessons learned have led to progress toward stronger and safer construction. In less well-developed regions, economics may prevent similar progress and leave residents in danger if earthquakes strike. In areas with less well-known earthquake risk, there may not be as much perceived need for resistant construction. In the Missouri, Arkansas, and Tennessee areas, for example, very large earthquakes do occur, but they are much less frequent than they are along the Pacific Rim. Earthquake provisions in the building code may take a lower priority than other requirements. Earthquake risk is now being taken very seriously in the Pacific Northwest, but the risk has been considered relatively moderate for so long that a huge number of possibly unsafe structures already exist.

Another area where seismological research has improved people's chances of surviving an earthquake is in the monitoring effort. Take the Southern California Seismic Network, operated by Caltech and the U.S. Geological Survey, as an example. The following paragraphs illustrate three past and future scenarios in which monitoring is shown to be an invaluable key to improving our response capabilities:

The year is 1952. A large earthquake occurs in the Kern County area, with heavy damage in Bakersfield, Tehachapi, and Arvin. Local police and fire personnel deal with the disaster using only their own resources. Later, the engineering and scientific communities try to learn something from the rubble, but there is no seismic network to provide information immediately after the quake. Most of the seismic stations are recording optically, on photographic paper (those trusty Wood-Andersons), and the seismograms have to be mailed to Caltech, developed, and interpreted by an experienced analyst before the quake and its aftershocks can be assigned locations and magnitudes. Any telephone call to the Seismological Laboratory, for information about the location of a quake, would likely be answered with a set of questions regarding the caller's location and how strongly the quake was felt. Any inquiries from the press involve reporters with notepads, and the public does not receive any information about the earthquake until the next day's newspaper.

The year is 1971. A large quake occurs in the northern San Fernando Valley. This time, the Seismological Laboratory can provide some information, since many of the seismic signals are continuously being sent directly to Caltech over leased telephone lines. The recording drums on the seismographs are themselves subject to shaking, however, so the seismograms are a little difficult to read. It is determined within an hour after the earthquake that the epicenter was in Newhall. Telephone reports from Newhall to the civil defense office indicate that the damage is not too bad. But, in fact, the earthquake was on a thrust fault that dips underneath the San Gabriel Mountains at a 45-degree angle. The epicenter may be in Newhall, but the fault has ruptured upward and to the south into the nearby towns of Sylmar and San Fernando. A busy freeway interchange is in shambles, as are two major hospitals. The phones are down, so nobody has reported the damage. The confusion costs heavy rescue operations precious hours.

The year is sometime in the future. The long-anticipated San Andreas rupture begins at Bombay Beach in the southeasternmost corner of California. "Smart" digital seismometers nearby immediately detect the strong ground shaking and send the information to two computers at Caltech (a primary and a backup system). Alarms are immediately sent out by radio to the emergency services agencies, utilities, railroads, and other subscriber groups. High-speed trains slow down. Elevators stop automatically at the nearest floor. Because the radio signal travels much faster than the seismic waves, all this happens several seconds before local residents feel any shaking. Meanwhile, the Caltech computers, now running on an emergency generator in the basement, continue to process the incoming seismograms. Within minutes after the earthquake, an intensity map, showing exactly where the strong shaking occurred, appears on computer screens at the emergency operations centers and on the Internet. Rescue personnel and repair crews from the gas and electric companies know exactly where to go to find the most serious problems. Aftershock probability estimates follow within the hour.

If the Bombay Beach quake were to happen tomorrow, all this hardware and software would not be ready, and there would be no early warning system. But the automatic identification and listing of an earthquake's location would happen. Those data are available now on the Internet. In fact, a seismologist is on duty twenty-four hours a day in southern California, with a laptop computer and cellular modem, ready to verify or fix any earthquake information the computers generate. Automatically generated instrumental (based on observed ground acceleration) intensity maps are available, as is the opportunity to fill out a questionnaire about your own earthquake experience for a more traditional Modified Mercalli intensity map. There is no guarantee that the Internet will be operating, of course, after the long-anticipated Big One, or that the various Caltech sites will be able to handle all the curious Web visitors. However, for small and moderate quakes, a greater amount of information is available much more quickly than it was even five years ago, during the Northridge quake. The early warning part of the system is under development, as part of a program called TriNet, an integrated effort through

Caltech, the U.S. Geological Survey, and the California Division of Mines and Geology. As part of TriNet, the seismic stations in southern California are currently being upgraded from passive sensors of ground motion to sophisticated computers unto themselves. These will replace all the computers and software that presently monitor data coming from those stations. Computerized integration of seismic information is one of the exciting waves of the future in seismology.

With regard to predicting and responding to earthquakes today, again it's a good news/bad news story. First the bad news: Earthquakes occur and will continue to occur in the future. Some of the quakes will be large, and many will occur in regions that are ill prepared. Scientists have yet to learn how to predict exactly when and where an earthquake will strike, so preparedness at all times is the wisest and safest track to take.

And now the good news: Our understanding of earthquakes, where they are most likely to occur, how often, and how severe the ground motion will be, has improved dramatically. Furthermore, the behavior of various types of manmade structures in an earth-shaking event is better understood than ever before. Where modern building codes are in force, weak, older, and unsafe structures are slowly being replaced by stronger, more earthquake-resistant construction. And in some earthquake-prone areas, efforts have been made to educate the public about the hazards and risks, and how to respond in an earthquake, particularly within school classrooms. See the list of readings and Web sites at the end of the book for more information on how to prepare for an earthquake.

We may never be able to predict exactly where and when a quake will strike. But with continued investment and efforts to learn more about the science and nature of earthquakes, we will be better able to warn of and prepare for their impacts. And finally, remembering that earthquakes are part and parcel with plate tectonics, consider for a moment what the Earth would be like without plate tectonics. There would be no mountains and possibly no life (if life originated at deep-sea vents, as some now think); the third planet from the sun would be a very dull place indeed.

Chapter 3

Volcanoes

Galeras Volcano, Colombia, 1993

In Colombia, more than 300,000 people live dangerously close to one of the region's most active and explosive volcanoes, Galeras. In January 1993, a meeting of volcanologists was convened to bring international experts together with Colombian scientists to study and discuss Galeras, and to help improve efforts to monitor its eruptive activity. As part of the meeting, field trips to the volcano were made to assess how it should be studied and what monitoring methods might work best. At the crater, scientists were able to collect samples and test or demonstrate the use of new instruments. The organizers of the meeting constantly checked with colleagues at the nearby volcano observatory to review the status of Galeras's activity. Everything appeared quiet, stable, and safe. Then the unexpected and unthinkable happened. Twelve people were on the volcano when rocks suddenly began to fall near the crater, and within seconds, Galeras unleashed a powerful and explosive eruption. Several of the scientists were killed instantly, while others were pelted with fiery rocks shot at high speed from the

crater. Those surviving the deadly blast suffered extremely serious injuries and were traumatized from witnessing the violent demise of their colleagues and friends. What had started as a meeting to help reduce the risks to the local population ended in tragedy for some of those hoping to provide that help. Today, research on volcanoes is moving toward the use of instruments that do not require scientists to go into the crater of an active volcano. Even so, the development and testing of such devices usually requires researchers to venture near or onto potentially active volcanoes.

Because no fiery red lava had poured from Galeras during the eruption, many people believed it had been a relatively benign blast. This could not be farther from the truth, and it illustrates one of society's most pervasive misunderstandings about volcanoes. Most people hear about volcanic disasters and associate them with pictures of glowing red lava shooting high into the air, or rivers of lava flowing down a barren black landscape. These are the awe-inspiring volcanic eruptions of Hawaii. They are extremely photogenic, and because they flow somewhat predictably, film crews are apt to use them for shows on volcanoes (*color plate 10*). In fact, no one has ever been killed directly by a Hawaiian lava flow. Most lava flows move so slowly that the average person can simply walk away from the danger. Houses and other structures are sometimes destroyed, but loss of life is rare. Volcanic eruptions take on a variety of styles, and Hawaiian-type lava flows are probably one of the least dangerous and least common of the bunch.

In 1980, the eruption of Mount St. Helens captured the attention of people around the world (*Figure 3.1*). It seemed

Figure 3.1
Eruption of Mount St. Helens, May 18, 1980. Courtesy U.S.
Geological Survey.

like an extraordinarily large and violent volcanic event. Huge
explosive blasts of hot gas and ash and massive torrents of mud
and volcanic debris surged off the erupting mountain. Yet the
Mount St. Helens eruption was really a rather ordinary event
by volcano standards. Volcanic outbursts on the same scale, in
terms of style, energy, and volume, occur at least once every
decade. And the explosive nature of the eruption was not
uncommon; in fact, most of the world's volcanoes have,
throughout history, erupted in a similar manner.

Today, roughly 500 million people live dangerously close
to the world's 1,500 active volcanoes. Fertile soil and majestic

volcanic peaks increasingly beckon both resident and tourist populations. Consequently, the number of people at risk from the dangers posed by potentially active volcanoes is rising. Our understanding of volcanoes and how they erupt has advanced greatly over the last thirty years, but it remains difficult to forecast specifically when and how eruptions will occur, particularly if no warning signs are detected. In the short term, forecasts are vastly improved, but over the long term they remain dubious. Recognizing these limitations, scientists are working throughout the world to better monitor, predict, and warn of impending eruptions so that in the future, loss of life and property can be averted.

Uncorking a Volcano

Many of the world's volcanoes are like huge bottles of champagne. Once uncorked, a violent and explosive blast of liquid and bubbly gas is released. Imagine an everyday bottle of champagne. It has thicker glass than an ordinary bottle, an oversized cork tightly wrapped by a wire cage, and a warning label cautioning people to use care while opening. Inside the bottle, the champagne contains a high concentration of dissolved carbon dioxide gas. When the bottle is sealed, pressure within it is high, so the gas remains in solution, as evidenced by the lack of bubbles. A tremendous amount of energy is bottled up, literally—enough to easily break a normal glass bottle. The instant the seal around the cork is broken, the internal pressure within the bottle plummets and bubbles rapidly begin to form, grow, and rise to the surface. This happens with such speed and energy that many a champagne cork has become a dangerous flying projectile. The eruption of a volcano can occur in much the same way.

Volcanoes begin where high temperatures deep in the Earth create a thick, semifluid molten rock, called magma. Magma, as explained earlier, is generally composed of dissolved gases (including water, carbon dioxide, and rotten egg-smelling sulfur dioxide) and melted or crystallized minerals. Once formed, magma tends to be less dense than the solid surrounding rock, so buoyancy drives it upward. Thus the magma rises, pushing or breaking its way through cracks and fissures in the Earth's mantle and crust, and sometimes collecting in large open spaces or chambers. As the magma moves closer to the surface, the surrounding pressure decreases, so gases within the magma begin to bubble out of solution (like the bubbles in the uncorked bottle of champagne). Because many gases are very insoluble in magma, they can begin to bubble out even at relatively great depths. When gas is released from the magma, it rises much faster than the molten rock below. If a vent or fissure is present at the surface, both gas and magma erupt. When magma erupts above ground, it is called lava. However, if a tight seal or rocky cork exists at a volcano's summit, the gas and magma will begin to accumulate dangerously beneath the surface. A swelling bulge or dome on the volcano might be indicative of the magma, gas, and energy building up below. If the overlying seal is suddenly broken, possibly by an earthquake or landslide, an explosive eruption occurs. The nature and size of the blast depends on the amount of gas, magma, and pressure released, as well as the chemical composition of the magma. But just as the champagne cork can fly violently through the air, so can volcanic rocks, hunks of lava, ash, and gas.

Tools of the Volcanology Trade

Unfortunately for modern volcanologists, early texts on volcanoes consist largely of beautiful photographs and qualitative descriptions of famous eruptions. While these books are fun to read and astonishing to look at, they provide little quantitative information about volcanoes. Over the last several decades, however, the development of new instrumentation has rapidly advanced our ability to measure and study volcanic activity. And modern computer technology allows for the efficient and fast manipulation of data, while remote sensing techniques provide for safer and larger-scale studies.

Much of modern volcanology involves the monitoring of volcanoes. Some scientists describe monitoring as keeping a detailed "diary" for each volcano, noting all of its daily changes, both visible and invisible. This may mean simply documenting observable variations in things such as plant life or the number of fractures and steaming vents along a volcano's slope. Monitoring might also entail the repeated sampling of a volcano's gas and magma, or measuring the changing shape of a volcano. One of the goals in volcano monitoring is to distinguish between everyday "noise" (variability) in the data and changes that signify an impending eruption. A wide variety of modern instruments are now used to study and monitor volcanoes, and they are providing a growing insight into their fiery underbellies as well as the clues that signify their growing unrest.

Seismographs are the old, well-established instruments of choice for the study of earthquakes, but they are also one of the most important tools of the volcanologist. Earthquakes and volcanic eruptions typically go hand in hand. Eruptions can trigger earthquakes, and earthquakes can trigger eruptions. Magma rising beneath a volcano must often break its way through the overlying rocks, thus creating earthquakes. These earthquakes can occur months, weeks, or days in advance of an eruption. Swarms of earthquakes almost always precede or accompany volcanic eruptions. Some of these swarms consist of numerous small but unusually long and rhythmic quakes, a phenomenon called volcanic tremor. Because they are vibrations released when a rock suddenly breaks, normal earthquakes are typically short, a few seconds to minutes at most. In contrast, long-period, rhythmic volcanic tremor seems to be caused by the movement of fluids or gas through cracks beneath the surface. It is like an organ's distinctive sound, created by pulses of air that are released into pipes of specific diameter and length. Because volcanic tremor is thought to result from the movement of molten rock or gas beneath the Earth's surface, when it is detected by monitoring instruments, scientists take note—magma may be on the move.

Just before his assassination, President John F. Kennedy signed the Anti-Ballistic Missile Treaty, and international politics furthered the science of volcanology. The treaty was possible because geophysicists had convinced President Kennedy that if there was an international array of standard seismographs, then the United States and Russia would be able to carefully monitor tests of nuclear weapons and missiles, anywhere in the world. So we all got a treaty, and scientists around the world got a network of seismic stations. Because earthquakes accompany volcanic activity, a widely distributed array of seismic stations enabled scientists to locate volcanically active regions throughout the world, even under the ocean.

On an individual volcano, a series of seismometers provide a means to directly monitor magma movement and furnish an unprecedented view of the volcano's internal plumbing. Scientists identify where magma movement is occurring by detecting earthquake activity and locating its origin precisely. On each volcano there must be an array of, at the minimum, three seismometers. Volcanologists also use the seismic data to interpret how much energy each earthquake releases. The number of quakes, their depth, precise origin, and the amount of energy released are all used to evaluate how active the volcano is and whether a body of magma is moving toward the surface.

In addition, with a concentrated array of seismic instruments and computer technology, similar to that used to image the human body with a CAT scan, it is now possible to produce computer-generated images of a volcano's interior. An array of seismometers placed around a volcano record seismic waves passing through the underlying rock. Because seismic waves are slowed, refracted, or inhibited by fluid or molten materials, seismic recordings reveal where there is magma and where there is solid rock. A computer then combines the seismic data and constructs a detailed image of the volcano's internal structure. Ideally, one can then "see" the underlying magma body and monitor its movement, particularly as it ascends before an eruption.

On the Big Island of Hawaii, the U.S. Geological Survey has operated a volcano observatory for more than seventy-five years. Here, provided with a natural laboratory conveniently equipped with long-term volcanic activity and frequent eruptions, scientists have been able to develop and test a wide variety of new instruments. For instance, researchers at the Hawaii Volcano Observatory have demonstrated that monitoring of a volcano's deformation can provide a powerful tool for forecasting the location, volume, and timing of eruptions. As magma and gas rise from beneath a volcano, the overlying rock may inflate, bulge, or shift ominously. And when an eruption actually begins, areas on a volcano will sometimes deflate or rapidly shrink. This deformation of a volcano's surface can be measured using recently developed and relatively inexpensive tiltmeters and laser technology. A tiltmeter simply measures the tilt or slope of the earth's surface. And highly precise, laser-equipped instruments measure the distance between points on a volcano.

To make laser distance measurements, scientists first choose specific points on a volcano's rocky surface and install special mirror reflectors. At a separate location, a solid, stable base is constructed for the installation of the laser instrument. The distance between the laser and reflector positions is then determined by the time it takes for the laser beam to travel from the instrument to the reflectors and back. If the laser beam's travel time changes between measurements, the rock on which the reflector sits has moved. These measurements, repeated over time, can reveal very precisely, on the order of one part per million, a volcano's changing shape. Conceptually, one part per million can be thought of as extremely weak chocolate milk— one drop of chocolate in sixteen gallons of milk. On Kilauea in Hawaii, deformation data shows that the volcano's surface tends to rise, or inflate, before eruptions. Given this information, and with the high quality of data obtained through monitoring, scientists at the Hawaii Observatory have been able to forecast the volume of eruptions, where they are most likely to occur, and when (by a few hours or more). But Hawaii's volcanoes are different from the majority of the world's eruptive peaks. Deformation and bulging do occur on other types of volcanoes, but so far deformation measurements alone have proved of limited use. In April 1980, a huge bulge grew on the northern side of Mount St. Helens prior to its eruption, but tiltmeters and laser distance monitors on its lower flanks failed to detect the extent of deformation. Unfortunately, distinct point measurements of deformation are limited by where the instruments are placed. On Mount St. Helens, they were in the wrong place at the wrong time. And not all volcanoes bulge before they burst.

Use of the Global Positioning System (GPS) is also facilitating exciting breakthroughs in the monitoring of volcanoes. GPS receiving stations can be rapidly deployed, take continuous readings, and provide virtually real-time and highly precise measurements of deformation. Satellite and laser altimetry, which measure the elevation of the Earth's surface from space or an aircraft, are promising, relatively new technologies that may provide even larger scale and more precise measurements. These types of remote and highly accurate measurements are undoubtedly the wave of the future in deformation measurements as well as in the monitoring of volcanic gas emissions.

The world's most explosive volcanic eruptions result from the sudden release of gas or gas-rich magma. Because gases rise through a volcano much faster than magma, an increase in emissions may reveal significant changes in a volcano long before an explosive eruption. For these reasons, great emphasis has recently been placed on developing tools for the safe and accurate measurement of volcanic gas emissions.

While studying ore deposits that form beneath volcanoes, Dr. Dick Stoiber at Dartmouth College became particularly interested in measuring volcanic gases. He thought that by sampling gas from an active volcano, one could directly observe the processes responsible for the creation of valuable ore deposits. In 1972, he pioneered the use of a remote sensing device to measure the flux of gas released from volcanoes. The instrument had previously been developed to measure sulfur dioxide gas in mineral smelters and coal-fired power plants. Sulfur dioxide is a high-temperature gas released from rising magma bodies, similar to carbon dioxide. The instrument Stoiber used is called the correlation spectrometer or COSPEC and uses a telescope to look into a volcano's plume of rising gas and measure the absorption of ultraviolet light. The absorption of ultraviolet light in the plume is a function of the amount of sulfur dioxide present. The COSPEC determines the amount of sulfur dioxide by comparing the change in ultraviolet light absorption in the volcano's gas with that of a standard gas inside the instrument. By driving or flying the device beneath a volcanic plume of gas, scientists are afforded a "snapshot" of the amount of sulfur dioxide being emitted. When this is multiplied by wind speed over the volcano, the rate of gas emission or flux can be calculated. And because no one has to directly sample gas or magma, it is a relatively safe procedure. The COSPEC is now considered one of the basic tools in volcanology.

Using the COSPEC, scientists found that volcanoes in Central America release, on average, approximately 200 to 500 metric tons of sulfur dioxide each day. In the days or weeks before a volcanic eruption, the rate of sulfur dioxide release rises to more than 1,000 tons per day. During eruptions, volcanoes can release millions of tons of sulfur dioxide, affecting the climate and producing significant environmental problems. High in the atmosphere, sulfur dioxide transforms into small particles of sulfuric acid. These particles reflect or absorb incoming energy from the sun, and if enough are present, can cause a cooling of the Earth's climate. When sulfur dioxide in the atmosphere combines with water vapor, acid rain is formed. Volcanically derived acid rain can damage nearby vegetation and crops.

Current research is focused on developing new and safer tools to monitor volcanic gases. Small, lightweight, and inexpensive instruments are being tested that could be deployed for weeks inside a volcano's crater. The goal is to measure carbon dioxide, sulfur dioxide, hydrochloric acid, temperature, and humidity in virtually real time. This would help researchers generate a long-term baseline or diary for individual volcanoes and detect significant and early changes in their behavior. Plans are to use simple radio transmitters to send data from an instrument to a remote base station so that fewer hazardous trips into an active crater are needed. Scientists also hope to compare variations in the chemistry and quantity of gas released with the record of earthquake activity to learn more about the relationship between the two.

New gas-measuring instruments have been tested at volcanoes in Colombia, Indonesia, and Russia, and while they are producing promising results, tests have also revealed numerous problems. Because a volcano's crater is an extremely acidic environment, many of the sensors and their internal computers are quickly ruined. Each type of gas being measured requires a specific sensor, and the new instruments also seem to have difficulty with the mixture of gases released by a volcano. So far, data suggest that accurate measurement of sulfur dioxide is possible, but determinations of carbon dioxide and hydrochloric acid remain somewhat imprecise. And as with most endeavors that combine the testing of new equipment and harsh environments, there have been plenty of other problems. Within a few days, acid gases destroyed solar panels used to recharge the automobile batteries, which provide power to the instruments. And lightning repeatedly struck the instruments. Given what has happened to the instruments, it is amazing that so many scientists and graduate students have survived going into volcanic craters to collect samples of gas and rock.

Progress is being made in determining carbon dioxide levels in volcanoes with a new device, named GASPEC, which measures the absorption of infrared light by molecules of carbon dioxide, similar to the absorption of ultraviolet light by sulfur dioxide. Carbon dioxide coming from a volcano is particularly difficult to measure, because volcanoes emit only slightly more carbon dioxide than is already present in the Earth's atmosphere. The atmosphere normally contains about 370 parts per million (ppm) carbon dioxide, and volcanic emissions typically add only about 5 to 15 ppm more. However, by comparing volcanic gas outputs with a reference gas inside the device, the GASPEC is able to precisely measure the relatively small, but highly significant amount of carbon dioxide coming from a volcano.

Carbon dioxide is very insoluble in magma, so as molten material rises beneath a volcano, it is released far in advance of other gases. Consequently, volcanoes may show significant increases in the flux of carbon dioxide long before (perhaps weeks or even months) rising magma arrives at depths shallow enough so that other gases, such as sulfur dioxide, are released. Scientists hope that by measuring the ratio of carbon dioxide to sulfur dioxide coming out of a volcano, they can track magma as it nears the surface; the closer it gets, the more likely an eruption.

Other instruments have been developed to measure gas coming from a volcano, but all involve direct sampling, either by going directly into the volcano or by flying repeatedly over it and through its acidic plume. Flying over an active volcano is not only dangerous, but very expensive and difficult, because the acidic plume can cause extensive damage to an airplane. Scientists at the U.S. Geological Survey and their Mexican colleagues have had success with instruments such as the LiCor. It has been used to measure carbon dioxide release from the huge and dangerously active volcano Popocatepetl, which sits precariously close to Mexico City and is being carefully monitored to detect signs of eruptive activity.

Satellite technology is providing another new and exciting tool in the study of volcanoes and gas emission. In the late 1970s, NASA launched a satellite to measure the Earth's ozone layer. In 1982, the scientist in charge (Dr. Arlin Kreuger) of the TOMS (Total Ozone Mapping Spectrometer) satellite observed that the data indicated an unusually large amount of ozone over southeastern Mexico. Dr. Kreuger discovered that a previously unknown volcano, El Chichon, had recently erupted and that the satellite sensor, specifically designed to measure the atmosphere's ozone concentration, was also measuring the UV absorption by sulfur dioxide in the upper atmosphere. The volcanic gases released by the eruption had created a false ozone reading on the satellite image, thus revealing the activity of El Chichon—a lucky break for volcanologists. Since that time, scientists have benefited a great deal from the TOMS satellite, using it to remotely quantify the sulfur dioxide released from many explosive eruptions (*color plate 11*). Other instruments on NASA's arsenal of satellites, as well as on the space shuttles, have also been used to study and document volcanic activity. The High Resolution Infrared Radiometer (HRIR) has been used to image changes in surface temperature associated with volcanic activity. And instruments on the Landsat satellites provide a long series of repetitive imagery that can be used to identify and map changes in landscape that result from eruptions, such as lava flows or ash deposits. The image of Mount Pinatubo shown on *color plate 12* is from a microwave radar instrument flown aboard the space shuttle *Endeavor* in April of 1994.

Another relatively new field in volcanology involves the tracking of a magma's source and generation through the study of geochemical tracers. By measuring trace elements or radiogenic and stable isotopes in lava and the surrounding or related rocks, scientists are attempting to track the genesis of magma at individual volcanoes. Some studies are also monitoring changes in the magnetic field around a volcano. Such changes have been noted prior to eruptions and may reflect the influence of temperature and/or magnetic minerals in the magma.

In addition to studying the present-day activity of volcanoes, scientists are also actively investigating their eruptive past. Just as sediments can record the past history of earthquakes (and tsunamis), so too can they tell of ancient volcanic eruptions. Whether through digging a trench or carefully examining outcrops of rock along a roadside, geologists can often identify sediment layers indicative of a particular eruption. The layer of volcanic sediment may have a distinctive color, mineral composition, or physical structure. Having identified an eruptive deposit, scientists document and map its extent and thickness throughout the region surrounding a volcano. The mapped layer is then used to interpret the size and type of eruption that created the deposit. The timing of the eruption can also be estimated, using radiocarbon or other dating techniques. If sediment layers suggest more than one previous eruption, then multiple events may be mapped and dated to also learn about the frequency of eruptions. However, sometimes exposing and interpreting the geology can be difficult, particularly if there is dense vegetation cover or if there has been extensive erosion. Weathering can reduce or alter the original volcanic deposit. GPS, geographic information systems, and computer mapping software are facilitating better, faster, and more accurate assessment of a volcano's eruptive history.

All told, methods for the study of volcanoes and potential volcanic activity have greatly improved over the last half century and are rapidly providing new and detailed information about the Earth's mountains of fire and their furious behavior.

Where Volcanoes Lurk

The concept of plate tectonics revolutionized our understanding of the Earth's dynamic nature, and it also provided a wealth of knowledge regarding volcanoes—where, how, and why they erupt. As described earlier, volcanism occurs most commonly at the angry edges of the tectonic plates, and at hot spots on their interiors (*Figure 3.2*). Volcanic activity comes in different styles, depending on the plate boundary type or hot spot at which the activity occurs.

At the plate's divergent borders, the mid-ocean ridges, we have never witnessed actual volcanic eruptions firsthand. Scientists in submersibles have found very fresh, warm fields of cooling lava on the sea floor, but no one has ever actually

Figure 3.2
Distribution of volcanoes that have been active over the last 10,000 years. Courtesy Smithsonian Institution, Global Volcanism Program.

seen an undersea eruption. However, evidence suggests that along the entire mid-ocean ridge system, more than 60,000 kilometers long, there are probably volcanic eruptions happening, somewhere, all the time. Seismic records from around the world indicate that small, shallow earthquakes occur often and in abundance along the mid-ocean ridges. These earthquakes may indicate where and when magma is erupting. Scientists are currently proposing to establish several remote (unmanned) undersea observatories along the mid-ocean ridge system that could automatically monitor and record sea floor eruptions.

Deep beneath the sea, at the mid-ocean ridges and elsewhere, pressure is high because of the great weight of the overlying water. When lava erupts on the sea floor, this high pressure keeps volcanic gas in solution, in much the same way that the thick glass and tight seal of a champagne bottle keeps the champagne from bubbling. Consequently, there are no explosive eruptions on the mid-ocean ridges or elsewhere in the deep ocean. Here, lava can only erupt in relatively passive flows and form rounded blobs and tubular pillowlike structures (refer to *Figure 1.5*).

All of the magma erupting at the mid-ocean ridges is similar in chemical composition and is collectively called basalt. Magma and volcanic rocks generally contain varying proportions of oxygen, silicon, aluminum, iron, calcium, magnesium, sodium, titanium, and potassium. Basalt contains up to 50 percent silica (SiO_2) and is rich in iron, calcium, and magnesium. The rock's mineral composition, particularly its iron, lends it a dark-gray to black color. Because so much basalt is extruded at the mid-ocean ridges, it is one of the most common rocks on Earth. Most scientists believe that basalt's composition is similar to that of the Earth's mantle. Through this constant churning of rock, essentially turning the Earth inside-out, volcanoes offer a rare window into the mysterious interior of our planet. Because lava erupting beneath the sea is cooled quickly, submarine basalts tend to be fine-grained and glassy. With rapid cooling there is no time for crystals to grow. In contrast, magma that cools more slowly generally creates rocks that are coarser grained and contain larger crystals. Granite, a familiar coarse-grained rock, is formed by the slow cooling of magma within the earth.

At the plate's convergent margins, volcanoes are commonly found in association with subduction zones. Linear mountain belts, such as the Andes, typically lie behind a subduction zone and the overlying deep-sea trenches. Topping off the mountains, or buried deep within them, are the volcanoes responsible for much of their bulk. The volcanoes form directly above the spot where a descending, subducted plate reaches a depth of about 150 kilometers. Scientists believe that at 150 kilometers down, a critical pressure and temperature point is reached at which hydrated minerals lose their water. The relatively buoyant magma then rises from the descending slab, triggers melting of the surrounding rock, and breaks through or deforms the overlying crust. Imagine a ribbon of molten rock lying beneath the Andes Mountains, perfectly aligned with their twisting chain. At regular intervals along the underground ribbon, blobs of magma rise and burst forth onto the land, where they form a necklace of regularly spaced volcanoes, doming up as mountains.

The erupted lava from volcanoes associated with subduction zones is quite variable in composition. Some of it is basalt, similar to that at the mid-ocean ridges, but a large proportion of the magma and the rock it creates is more enriched in silica. These rocks include andesite, dacite, and rhyolite, which in comparison to basalt contain increasingly higher amounts of silica. And rocks with a greater amount of silica also tend to have comparatively more sodium and potassium, and less iron, calcium, and magnesium. Because of these differences, andesite tends to be medium gray to gray in color, while dacite and rhyolite are light gray to tan. The process by which magmas evolve from a basaltic or primitive composition to become andesite, dacite, and then rhyolite is known as differentiation, and it occurs when the magma rises toward the surface. As magma ascends, it cools and is exposed to lower pressures and temperatures. As a result, some minerals begin to crystallize. Different minerals crystallize at different rates, and at varying temperatures and pressures. So depending on the rate and nature of the process, a variety of magma mixtures can form. The molten rock rising from a subducted slab may also begin to crystallize at the same time that it is melting and mixing with the material from overlying rocks. Behind a subduction zone the overlying rocks tend to be variable in composition, so with all that mixing, a wide variety of magma types are created.

The Hawaiian Islands represent a third and very different plate tectonic setting for volcanism. For reasons that are still unknown, there are "hot spots" deep in the mantle where unusual chemical compositions and temperatures exist. At a hot spot, as described earlier, magma rising through the mantle essentially burns and breaks its way through the overlying tectonic plate. Hot spots can be large and stable for long periods, even many millions of years. For instance, the Hawaiian Island chain formed, and is still forming, as the Pacific Plate passes over a hot spot that has been fixed in place for at least 70 million years (refer to *Figure 1.8*). This chain of some 137 islands stretches northwest all the way across the Pacific Ocean. At the far northwestern edge of the chain, the Pacific Plate is being subducted beneath the Russian peninsula of Kamchatka —site of some of the largest and most active volcanoes in the world. The Big Island of Hawaii currently exhibits the most active volcanism in the chain, although volcanism has also recently occurred on the island of Maui. And beneath the dark waters of the sea, a new island, 28 kilometers southeast of the Big Island, is beginning its rise to the surface. The top of the fledgling seamount, named Loihi, currently lies some 970 meters below sea level. Observations from submersibles suggest that it is covered with young pillow basalts (*Figure 1.5*), and that the overlying water is warm and rich in carbon dioxide.

The Big Island of Hawaii is the site of five volcanoes considered potentially active, one of which, Kilauea, has been in a state of almost constant eruption since 1983. Mauna Loa is the largest volcano in the world, and it looms over Kilauea. Reaching higher than 4,000 meters (13,000 feet) above sea level, its majestic crown is often covered with a blanket of white snow. For much of early 1984, Mauna Loa and Kilauea exhibited simultaneous eruptions, leading some to speculate on an underground connection between the two. Scientists now believe that the magma chambers beneath Kilauea and Mauna Loa are not connected, but are completely separate entities.

Surprisingly, Hawaiian volcanism is somewhat unique in a number of ways. The rocks are always of basaltic composition, and the eruptions are never highly explosive. Beautiful streams and fountains of fiery, fluid lava burst from the Earth and create glowing rivers that flow for great distances over the blackened landscapes (*color plates 10 and 13*). Another site of this unusual Hawaiian-type volcanism is in the Indian Ocean, on the French Island of Reunion.

There is one other plate-tectonic setting that exhibits volcanism. At the Great African Rift Valley in Eastern Africa, spanning the countries of Ethiopia, Kenya, Tanzania, Rwanda, and Congo, the ancient and very stable continental crust is, for uncertain reasons, breaking apart. Here, the African plate is literally being ripped apart from below and a new zone of plate divergence is forming, heralded by the presence of active volcanoes. Ultimately, if the edges of the plate continue to separate, a new mid-ocean ridge and basin will form in the central rift valley.

The volcanoes of Eastern Africa are of widely variable and sometimes exotic composition. In the creation of a new rift, magma from the Earth's mantle rises through ancient and highly evolved rocks. In the process, some of the surrounding rocks are melted and added to the rising molten material, resulting in a magma (and ultimately a lava) of complex composition. The region includes many explosive volcanoes, including Oldoinyo Lengai. Here, the source of magma is very poorly understood, and lava coming from the volcano is unusually low in silica and high in carbonate. The lava flows are thin and move as easily as water at remarkably low temperatures (as low as 600° C, compared to typical lava temperatures of over 1,000° C).

The study of plate tectonics has revealed why volcanoes do not occur randomly over the Earth's surface, but in neat, linear chains, belts, or ridges. And it begins to explain why the nature of volcanoes varies.

Types of Volcanoes

Volcanoes are landforms built of eruption. Consequently, the style and type of typical eruptions determines the shape and style of volcano. As noted earlier, most people think of volcanoes in terms of the Hawaiian stereotype—broad, blackened domes that ooze and spit fiery fountains and rivers of orange-red lava. But in fact, these types of volcanoes are a minority. Hawaii's mountains of fire are called shield volcanoes, because they have a broad, gently sloping shape somewhat reminiscent of a warrior's shield. They form as successive flows of liquid lava pour out of the summit, cool, harden, and pile up into a wide, domelike structure. The lava flows typically are highly fluid and basaltic in composition, and they spread out over long distances. As these fiery torrents of lava flow down a volcano's side, they cool to form gently dipping sheets of crusty black basalt. Few explosive eruptions occur, because the fluidity of the lava allows gas to escape easily and there's no cork stopping the flow. At the summit, many shield volcanoes have a large circular depression called a caldera. Most calderas form when erupting magma empties an underlying chamber, and the overlying rock collapses. Smaller collapse or explosion features are called craters. Sometimes magma also erupts from rift zones, fractures, and fissures along a volcano's side.

Probably the most famous and recognizable shield volcanoes are Hawaii's Mauna Loa and Kilauea. Spectacular rivers of fiery lava pour from these volcanoes and are ever building massive mountains of black basalt. Over the last two centuries, both Mauna Loa and Kilauea have erupted, on average, every two or three years. Some of Mauna Loa's rivers of fire have been clocked at 56 kilometers per hour, but most flow steadily at a slower pace. Occasionally, lava will shoot skyward in magnificent incandescent fountains. During a 1959 eruption on Kilauea, a glowing fountain of lava reached a height of some 580 meters (1,900 feet). Photographs of Hawaii's lava flows and fountains are undoubtedly some of the most spectacular images ever captured of the furious Earth at work (*color plates 10 and 13*).

As lava flows down the slopes of a shield volcano, it begins to cool. The front edge of the flow often starts to harden, only to be overtopped or pushed on by the flaming torrent behind. Sometimes the edges and top surface of the lava may cool and harden while the interior portion of the flow remains fluid and blazingly hot. At the sides, hardening crusts can form natural levees that channel the flow. And if the sides and top of a flow harden, a lava tube can be created. The interior of the flow then becomes insulated from the cooling effects of the air and may continue to flow for great distances. During eruptions of Kilauea in the early 1970s, a tube system was formed that funneled lava flows some 11 kilometers to the sea. In volcanic regions, such as the Galapagos Islands, ancient lava tubes have been preserved as long-running hollow caverns that make for fascinating, although dark, hikes. Where incandescent lava plunges over a steep cliff, a glowing lava fall may form (*color plate 13*). And where molten rock flows into a preexisting crater, a lava lake is created. On a lava lake, a thin black crust may form over the lake's surface, but the flow and sloshing of the liquid below often creates cracks and fissures in the overlying roof. In both lava lakes and tubes, a hardened black surface hides the fiery, still-molten rock that flows below. With extreme care, researchers can use either natural or forced holes in the thin roof of a lava flow or lake as a window for sampling.

In Hawaii, liquid lava flows and the rocks they form are generally described as either *aa* or *pahoehoe* flows. *Aa* lava is a thick, viscous fluid that cools to create rough, blocky and fragmented rock (*Figure 3.3*). Many a researcher's shoes have been worn away by the unforgiving sharp terrain of an *aa* flow.

Figure 3.3
Aa lava flow. Courtesy U.S. Geological Survey.

In contrast, *pahoehoe* flows are thinner and less viscous and form relatively smooth, ropy rocks. Cooled *pahoehoe* flows

Figure 3.4
Pahoehoe lava flow. Courtesy U.S. Geological Survey.

literally freeze into the rock swirls, twists, and blobs of a lava flow (*Figure 3.4*). Often the rocks and crusts formed from cooling lava are riddled with holes, the remnants of bubbles in the liquid lava.

Where hot lava meets the cold sea, towers of steam rise high in the air. Liquid lava that falls into the ocean is cooled extremely fast and often shatters into small, shiny pieces of black glass—creating a shoreline blanketed with black sand. The process is similar to what happens when cold water is poured into a hot glass coffeepot: say goodbye to the pot, because it will almost certainly crack.

In contrast to the broad dome of a shield volcano, most volcanoes have steep-sided, symmetrical peaks with relatively smaller craters at their summit. These are called either composite or stratovolcanoes; they are formed by the pileup of alternating layers of lava and pyroclastic debris. *Pyroclastic* or *pyroclasts* refers to fragments, large or small, that are ejected out of a volcano. They range in size from tiny ash particles, to larger cinders, and even bigger bombs and blocks. Pyroclastic debris can mix with hot gas and form explosive, fast, searing flows sometimes referred to as a *nueé ardente*, French for "glowing cloud." When a composite volcano erupts, it can be one of the most powerful and deadly types of explosions on Earth. Some of the world's most majestic and dangerous peaks are composite volcanoes, including Japan's Mount Fuji, Italy's Mount Vesuvius, Indonesia's Krakatau, California's Mount Shasta, Oregon's Mount Hood, and Washington's Mount St. Helens and Mount Rainier (*Figure 3.5*).

Composite volcanoes typically begin an eruptive sequence with an explosive blast of ejected debris and gas. After most of

Figure 3.5
Mount Rainier, Washington. Courtesy U.S. Geological
Survey.

the gas and gas-rich magma has been emitted, thick, viscous
lava is discharged. Over time, with repeated eruptions, alternating layers of pyroclastic debris and hardened lava build up
to form the volcano's core. At their summits, most composite
volcanoes have a circular crater with a central vent. Magma
tends to flow through a conduit or tunnel beneath the volcano
and erupt through its central vent or from fissures lining its
sides. When the fissures around the cone of a volcano become
filled with hardened lava, it provides support and strength
like the ribs of a boat. If the volcano's central vent becomes
plugged, a secondary vent may open and create a small adjacent cone.

The volcanoes that lie behind subduction zones are composite cones. The andesitic or rhyolitic lava that flows from
these volcanoes is thicker and more viscous than the basaltic
flows of a shield volcano. Thus, they tend to have steeper sides,

and rocky, jagged slopes. During explosive eruptions of composite volcanoes, a light-colored rock riddled with bubbles of gas is often created. This is known as pumice and typically contains so much trapped gas that it floats in water.

Composite volcanoes that have not erupted for a long time are considered dormant. They may lie quiet for centuries. While a volcano sleeps, erosive forces wear down the summit, softening its contours and often lulling nearby residents into a false sense of security. Sometimes magma that has hardened inside a volcano is more resistant to weathering than the surrounding rock, so eventually all that may remain of the original volcano is a hardened plug or neck of volcanic rock. Wyoming's rock formation named The Devil's Tower, which juts out high above the surrounding plain (*Figure 3.6*), is a solidified volcanic neck, a remnant of a volcanic cone worn away by the forces of time and nature.

Sometimes lava is so viscous that it neither flows nor is ejected skyward but rather bulges upward to form what is known as a lava dome. Lava domes essentially swell from inside, causing cracking and splintering of rock into fragments that roll and tumble down their sides. Domes commonly occur on the sides of composite volcanoes, and their growth or deformation may be monitored as an indicator of volcanic activity. Two of the most famous examples of dome growth that preceded powerful and explosive eruptions occurred with the deadly 1902 blast of Mount Pelee on the Caribbean island of Martinique and the 1980 eruption of Mount St. Helens in Washington. Eyewitness accounts suggest that a bulge grew on Mount Pelee just before a powerful flow of burning pyroclastic ash and gas erupted and killed all but two of the 30,000

Figure 3.6
The volcanic neck known as the Devil's Tower, Wyoming.
Courtesy U.S. Geological Survey.

residents in the nearby town of St. Pierre. And prior to the
eruption of Mount St. Helens in 1980, a bulge on its northern
face swelled some 137 meters (450 feet) high and 30 meters (85
feet) wide. It was the eventual collapse of the dome on Mount
St. Helens's northern flank that triggered the devastating blast

of May 18, 1990, the details of which are discussed later in this section.

Probably the most common and simplest volcanic structure is a cinder cone. As the name implies, a cinder cone is simply a cone built of cinders ejected from a volcanic vent. Gas within the erupting magma causes lava to be blown violently into the air, where it fragments into large particles and fiery blobs. These cinders cool, solidify, and fall to the ground around the erupting vent and create a circular cone of volcanic debris. In the early 1940s, a farmer in Mexico received a rude reminder of the Earth's furious nature when a volcanic eruption unexpectedly began right in the middle of his cornfield. There had always been a mysterious hole in the farmer's field, which he had used as a convenient place to dump rubbish. In late February 1943, the ground beneath his field began to shake and rumble, and a large crack opened across the strange hole. Much to the farmer's astonishment and terror, within days a cinder cone some 9 meters tall had formed and was belching fire and ash. After a week it stood at 150 meters, and one year later, what was once a flat field of corn had become a steaming cinder cone 410 meters (1100 feet) high.

Highly fluid, basalt lavas can also erupt through long fissures in the Earth. When this occurs, large, flat plains or plateaus, called flood basalts, are created. Extensive flood basalts form broad, low-lying plateaus in Iceland and near the Colombia River in Washington and Oregon. The Ocean Drilling Program recently focused its attention on a large, mysterious undersea area of flood basalts in the southern Indian Ocean. Scientists now believe that the flood basalts of this Kerguelen Plateau erupted 110 to 90 million years ago. And although the eruptions may have been similar in style to those on Iceland and in Hawaii, data suggest that the scale of the event may have been greater than any other in recorded history. Surprisingly, samples drilled from the submarine plateau also contained wood fragments, plant remains, and continental rocks. Researchers now think that at one time in its history, the surface of the Kerguelen Plateau was above sea level, and that it may contain useful information about the breakup of Australia, India, and Antarctica some 130 million years ago.

Some rare volcanoes have lakes atop their summits. These can only form where a special balance exists between an underlying source of heat and an overlying source of water. Episodic and violent releases of toxic gas trapped at the bottom of a volcanic lake can have deadly consequences, as in the volcanic event of Lake Nyos in Cameroon, which resulted in the deaths of 1,700 people. This event is discussed in more depth later on.

How Big Is Big?

An earthquake's size is described in terms of its magnitude, and the Richter scale or moment magnitude is used to compare the size of one event to another. Volcanologists have searched for a similar means to describe the relative size of volcanic eruptions. A variety of schemes have been suggested based on factors such as the height of an eruption column, the volume of the erupted lava and debris, the quantity, distance, and height of ejecta, and the duration of an eruption. The Volcanic Explosivity Index (VEI) is one scale that has been proposed. It ranges from a nonexplosive, gentle eruption of liquid lava,

VEI 0, to a violent, colossal explosion lasting more than 12 hours, VEI 8. An explosive eruption is considered very large or cataclysmic if it is a VEI 5 or higher. Eruptions of this magnitude tend to occur on average only once every ten to twenty years. Through historical accounts and geologic deposits, scientists have estimated the VEI for more than 5,000 eruptions in the last 10,000 years, and none has received a VEI of 8.

As with earthquakes, where magnitude does not necessarily correspond to the amount of damage done, the impact of a volcanic eruption may not be evident from its VEI ranking. Volcanic eruptions with a high VEI may do little damage, and those with a low index may be catastrophic in terms of deaths and economic losses. For example, the Nevado Del Ruiz eruption in November 1985 killed 23,000 people, yet the VEI is given at 3. The Mount St. Helens eruption was ranked as a VEI 5 but killed only fifty-seven people. The size or explosive nature of an eruption only partially determines the number of resulting casualties; other influential factors include how well and how far in advance the blast is forecast, if at all, and how well prepared communities are to respond to the threat of a volcanic eruption, as well as the location of the blast in relation to nearby populations.

Warnings, Precursors, and Eruptive Styles

As can be expected, the more scientists study volcanic eruptions, the better they understand volcanoes. With this knowledge, however, also comes a better understanding of what they don't know. One thing is for sure when it comes to volcanoes: the unexpected can and does happen. Deadly eruptions can come from volcanoes believed to be extinct—or from those that have not even been recognized as volcanoes. For example, in 1902, Santa Maria (Guatemala) produced a major eruption, although the volcano was supposedly extinct; and in 1968, Arenal (Costa Rica) was thought to be simply a quiet mountain peak until it unexpectedly erupted and killed seventy-five people.

Some volcanic eruptions exhibit similarities, but many are strikingly different. Volcanoes, one might argue, have unique personalities and must be individually studied to determine what kind of eruptive activity is likely and how often it can be expected to occur. Will a volcano erupt explosively, or is it characterized by slow lava flows? How big will the eruption be, and how large of an area will be affected? For volcanoes that have erupted in the recent past, we may know what types of eruptions are probable, but for those that have long been dormant or are literally unknown, this is more difficult to assess.

Italy, 79 A.D.: Vesuvius

One of the worst volcanic disasters ever recorded struck in the summer of 79 A.D. along the fertile shores of the Bay of Naples, Italy. Eyewitness accounts and the region's geology tell of a massive blast that decimated the land and its people. At the time, the prosperous cities of Pompeii and Herculaneum lay at the foot of a large, peaceful mountain called Mount Somma or Vesuvius. Although few in either city may have realized it, the rich soil that produced their lush crops and fruitful vineyards came from the nutrient-laden ash deposited during past eruptions of the nearby volcano. For years Vesuvius slept quietly.

Then on August 24, 79 A.D., the mountain came back to life with a vengeance. Reportedly, a few quakes occurred days before the eruption, but even if they had been recognized as precursors, no one would have imagined what was to follow.

The eruption of Vesuvius began with a blast of gas and ash that created a pine-tree-shaped cloud rising high into the sky. The initial blast lasted for a period of some eleven hours, and the cloud may have reached a height of 19 kilometers. Ash, rocky debris, and fresh, glassy pumice were ejected into the atmosphere, where prevailing winds carried them toward Pompeii. Reports suggest that cinders, ash, and pumice began to rain down on the city's inhabitants, accumulating at a rate of 15 centimeters per hour. Day turned to night as ash blocked out the sunlight and Pompeii's terrified residents scrambled for cover. Upwind of the volcano, the residents of Herculaneum saw little more than a light sprinkling of ash.

By nightfall, the volcano had belched out most of the gas-rich magma that had fueled its towering ash cloud, and the enormous mass of ash and gas hanging above its summit collapsed. The collapse triggered huge pyroclastic flows, burning torrents of hot ash and gas that surged from the volcano's summit and rolled through Herculaneum, incinerating all in their path. Throughout the night, repeated flows surged through the town and buried it under a burning shroud of volcanic debris. At first Pompeii was protected by its massive city walls, which diverted the burning plumes of gas and ash. But throughout the night and into the early morning, flows continued to cascade off the volcano and blaze through the countryside. Eventually, Pompeii too fell victim to the unleashed power of Vesuvius. A raging, swirling storm of red-hot ash and gas blew through Pompeii, destroying most of its buildings and entombing its fleeing residents in volcanic debris. Flames and dark roiling clouds of ash and gas could be seen for great distances across the Bay of Naples. Earthquakes shook the land, and a swelling tsunami sucked the water out of the region's harbors.

When the eruption finally abated and a murky daylight returned, the region was buried in deep drifts of ash. It is believed that 25,000 people perished during the hellish eruption of Vesuvius. Pompeii lay buried by 2.7 meters of pumice, topped by 2 to 3 meters (6 to 10 feet) of pyroclastic debris. Where Herculaneum once lay, there existed only a barren, dark volcanic deposit some 18 meters (60 feet) deep. Both cities remained lost until 1709, when some of their remains were uncovered by laborers digging a well. In Pompeii, archeologists uncovered hollow forms within the thick ash that were the molds of those caught in the eruption, their bodies having decayed long ago. Plaster casts made of the molds picture, in horrific detail, those who perished that day.

Vesuvius has erupted intermittently since the destruction of Pompeii and Herculaneum. In December 1631, 4,000 people were killed when the volcano unleashed another series of deadly pyroclastic flows, and in 1944, another eruption ejected ash and gas skyward, sending lava and mud flows racing toward the sea. Since then, Vesuvius has remained ominously quiet. Based on its past history, scientists believe its eruptions exhibit a distinctive pattern. Following a major eruptive event, Vesuvius becomes quiet for several years. Mild volcanic activity occurs as rising magma creates a new cone and fills the volcano's crater. When the pressure below builds high enough and the "cork" is unsealed, another colossal explosion occurs and the cycle begins again.

Eruptions exhibiting characteristics like the Vesuvius 79 A.D. event are now called *Vesuvian* and tend to include the expulsion of great quantities of ash-laden gas that discharge violently and form a high-reaching cauliflower- or mushroom-shaped cloud. One of the best eyewitness accounts of Vesuvius's eruption comes from Pliny the Younger, an eighteen-year-old from a neighboring island whose father, Pliny the Elder, was killed in the event. The most powerful type of volcanic eruptions are now called *plinian*, and these usually include a forceful ejection of relatively viscous lava and fast-moving pyroclastic flows. In contrast, nonexplosive eruptions of more fluid lava are referred to as *Hawaiian*-style events.

Washington State, 1980: Mount St. Helens

During March 1980, a sleeping giant in southwestern Washington named Mount St. Helens awoke from a 123-year hibernation (*Figure 3.7*). Beginning in mid-March, swarms of earthquakes announced the restless stirrings of magma and gas beneath the summit. For a week and a half, earthquake activity increased, and small avalanches of snow and ice cascaded off the mountain's peak and slopes. On March 27, the first eruption occurred. It was a thunderous blast of steam and ash that rose nearly 2,000 meters high and created a 76-meter-wide crater with fractures across Mount St. Helens's peak. During the next two weeks, the volcano intermittently spewed ash and steam, and a second slightly larger crater joined the first. Avalanches of snow, ice, and ash created dark streaks running

Figure 3.7
Mount St. Helens, Washington, before May 1980 eruption.
Courtesy U.S. Geological Survey.

down the mountain's slopes. During late March, seismic stations began to record rhythmic ground-shaking, or volcanic
tremor, possibly an indication that magma and gas beneath the
summit were on the move.

In late April and early May 1980, Mount St. Helens
became eerily quiet as all visible signs of volcanic activity
ceased. Then, between May 7 and 16, small intermittent blasts
of steam occurred, and a bulge grew on the mountain's northern flank. Through mid-May, some 10,000 earthquakes

occurred, many of which originated directly beneath the swelling bulge. And on the summit, directly behind the bulge, the mountain peak began to deflate. The volcano's deformation during this time was thought to reflect the movement of magma from deep in the Earth to just beneath the mountain's swelling northern flank.

Sunday, May 18, 1980, began bright and clear, like most days high on Mount St. Helens. Volcanologist David Johnston was working for the U.S. Geological Survey at an observation post nearly 10 kilometers north of the mountain. Measurements of seismic activity, gas emissions, ground temperature, and deformation that morning suggested that the volcano's activity was as it had been for the past month or so—Johnston reported nothing unusual. Then, just after 8:30 A.M., a magnitude 5.1 earthquake rocked the summit. In the next few moments, the bulge on the volcano's northern flank collapsed and unleashed a series of powerful eruptive events that would kill fifty-seven people, including Johnston.

Geologists Keith and Dorothy Stoffel happened to be flying in a small plane over Mount St. Helens on the morning of May 18. Their eyewitness accounts suggest that the whole north side of the volcano gave way almost instantaneously. The entire mass seemed to ripple, churn, and then slide to the north. A huge explosion then ripped through the summit, appearing to shoot out from beneath the sliding bulge (refer to *Figure 3.1*). The pilot of their plane made a steep dive to gain speed and outrun a growing massive cloud of ash and gas. By now, an enormous avalanche of volcanic debris was cascading off the volcano's slopes, moving north at speeds estimated at 248 to 288 kilometers per hour (155 to 180 mph). Like the uncorking of the champagne bottle described earlier, the sudden unloading of rock from the volcano's peak released pent-up gas and gas-rich magma from beneath the summit. An enormous pyroclastic blast of ash, rock, and hot gas surged laterally, at about 960 to 1120 kilometers per hour, off Mount St. Helens toward the north, followed by a nearly vertical explosion of ash and gas. This spread up and out, forming a mushroom-shaped cloud that within less than ten minutes had reached a height of some 19 kilometers. Forest fires were ignited by lightning strikes generated from the swirling clouds of gas and ash over the volcano. The spreading ash cloud blew into the nearby towns of Yakima and Spokane, Washington, and turned day into night. In the town of Ritzville in eastern Washington, nearly 5 centimeters of ash blanketed the ground. And in the upper atmosphere, carried by high-altitude winds, ash spread around the globe.

But the eruption was far from over. Soon new pyroclastic debris flows surged northward off the volcano in a boiling, frothy mixture of gas and magmatic debris—one scientist likened it to a pot of oatmeal boiling over. These flows were so hot that even two weeks after the event they were at a temperature of 300° to 418° C. As burning flows avalanched off the volcano, water encountered along the way instantaneously turned to steam and shot skyward. Searing gas and ash flows scoured the mountainside, rapidly melting glacial ice and snow. Soon huge volcanic mud flows, often called lahars, swept off Mount St. Helens, choking rivers and inundating the region with sediment and debris. The mud flows surged through valleys and over 70-meter-high hills, pushing the Toutle River 6.4 meters above its normal level.

In all, the 1980 eruption of Mount St. Helens lowered the elevation of the volcano by more than 300 meters and left a gaping crater 1.6 kilometers wide, 3.2 kilometers long, and 640 meters deep on its summit (*Figure 3.8*). Most of the volcano's victims died from inhalation of hot volcanic gas and associated injuries. The lateral blast, debris avalanche, and mud flows caused extensive damage to land and property. Thousands of acres of forest, recreational sites, bridges, trails, and houses were completely destroyed or heavily damaged. Fifteen miles

Figure 3.8
Mount St. Helens, Washington, after May 1980 eruption.
Courtesy U.S. Geological Survey.

of railway, as well as 296 kilometers of highway and roads, were buried or destroyed. Animals, birds, and fish throughout the region perished, and downwind of the volcano many crops were ruined. It is estimated that the Mount St. Helens eruption cost nearly $1.1 billion in losses. (The information contained in this account of the 1980 Mount St. Helens eruption comes principally from an excellent booklet by Tilling et al., published by the U.S. Geological Survey; see the list at the end of this book for more information on this and other excellent USGS references.)

Following the devastating eruption, researchers from the U.S. Geological Survey, the University of Washington, and other groups went to the site to further investigate the event. Much like surveys done following an earthquake or tsunami, observations made directly after a volcanic eruption provide invaluable insight into the particular event and can help forecast and plan for eruptions in the future. In addition, by looking at the deposits from a modern volcanic eruption, geologists are better able to interpret similar sediments laid down during ancient events.

Some of the most dramatic evidence of the impact of Mount St. Helens's 1980 eruption came from the area directly affected by the nearly supersonic lateral debris blast. Scientists divided the direct impact region into three zones radiating out from the volcano's northern flank. The innermost region, about 12.8 kilometers in radius, was called the direct blast or tree-removal zone. Here, the searing heat and power of the blast either burned or swept virtually everything away. The angle of charred and splintered tree remains was used to confirm the direction of the blast. The next area, extending out to a distance of about 30 kilometers from the volcano, was the tree-down zone. In this area, the force of the blast flattened everything in its path. Like a mowed lawn or pile of aligned matchsticks, trees lay on the ground snapped at their trunks, pointing in the direction of flow (*Figure 3.9*). The outermost area was called the seared or standing dead zone. Here many of the trees remained standing but were little more than bare upright sticks, burned dark by the hot gases of the blast.

Following the eruption, scientists also mapped the sediment deposits created by the eruption's massive debris avalanches and mud flows. Much of the debris flow had traveled down the North Fork Toutle River, scouring the valley and filling it with volcanic and glacial debris. A bumpy deposit topped with debris

Figure 3.9
Trees mowed down by Mount St. Helens lateral blast.
Courtesy U.S. Geological Survey.

mounds was found to extend downriver some 21 kilometers, with an average thickness of about 50 meters. The avalanche of debris had also flowed into Spirit Lake, creating a tsunami and filling the lake with almost 90 meters of material. Throughout the region, mud flows left a dirty "bathtub ring" of mud and debris in their wake. Although the mud flows left relatively thin deposits, averaging less than 1 meter thick, in some regions evidence suggests that at their peak, they were 10 to 20 meters deep. Mapping also revealed that at least seventeen separate pyroclastic flows had occurred, creating a fanlike pattern of overlapping sheet and lobed deposits.

Data from the 1980 event have also been compared with the geologic remains of past eruptions. For instance, old deposits suggest that a similar, although smaller, lateral blast occurred about 1,100 years ago. Evidence further suggests that mud flows, dome building, and pyroclastic blasts have also occurred periodically since Mount St. Helens first began to form some 50,000 years ago. In fact, geologists had been studying the eruptive history of Mount St. Helens for years prior to the 1980 event and had voiced concern about its potential for volcanic activity. In 1968, the *Christian Science Monitor* reported one geologist as being "especially worried about snow-covered Mount St. Helens," and in the late 1970s several geologists warned of an impending eruption, possibly before the turn of the century (Crandell and Mullineaux, 1978).

Since the catastrophic eruption in 1980, Mount St. Helens has exhibited intermittent and relatively benign volcanic activity. Most events have entailed the extrusion of stiff, viscous lava inside the volcano's crater. If it is correct to assume that Mount St. Helens will behave as it has in the past, then it would appear that for now, another major explosive blast is unlikely to occur in the very near future. Mount St. Helens seems to alternate between periods of eruptive activity and quiet dormancy. During the past 40,000 to 50,000 years, the volcano has exhibited nine active intervals. We are in the midst of an eruptive interval, and scientists expect that lava flows, doming, and small spurts of gas and ash are likely to occur intermittently until this volcanic giant returns to its next period of dormant sleep. Scientists today continue to closely monitor Mount St. Helens.

Like the deadly blast of Vesuvius, the 1980 Mount St. Helens eruption is considered a highly explosive and violent plinian eruption, yet there were only fifty-seven fatalities. Here we begin to see the benefits of modern volcanology. Nearly two months before the main eruption, a 4.2-magnitude earthquake alerted seismologists from the University of Washington and the U.S. Geological Survey, and they began intense, round-the-clock monitoring of Mount St. Helens. As the earthquake activity continued to mount, scientists discussed its significance and the potential hazards with local officials. Dr. Donal Mullineaux of the U.S. Geological Survey, considered one of the foremost experts on Mount St. Helens, was called in to review the situation and discuss the types of eruption that were possible and the hazards they posed to the surrounding region. Soon thereafter, local officials closed off areas around the volcano, even beyond its immediate slopes. On March 27, an official Hazard Watch was issued for Mount St. Helens, and a little over four hours later the first eruption of Mount St. Helens in the last century began.

By the second week of eruptive activity, twenty-five to thirty scientists were on hand to monitor the volcano and assess the potential hazards. Daily meetings and briefings were held with local officials, and on April 1 a large-scale volcanic-hazards map was released. During late April, scientists were aware that a collapse of the rapidly forming bulge on Mount St. Helens's northern flank could trigger an eruption, but no one knew for sure if and when it would occur. And even if it the bulge did fail, they could only guess how big the resulting eruption would be or what it would involve. Based on the increasing threat of an eruption, officials fortunately took measures to close off access to further areas at risk surrounding the volcano. Without such measures, the explosive eruption in May could have been more deadly, more like that of Vesuvius.

In 1982, the U.S. Geological Survey office responsible for monitoring Mount St. Helens and other volcanoes in the Cascades Range was designated the David J. Johnston Cascades Volcano Observatory. Like Hawaii's Volcano Observatory and others in Alaska, California, and throughout the world, scientists at the Cascades facility continue to provide valuable information on volcanoes through a wide variety of research, monitoring, and management activities.

Since the 1980 event, scientists have successfully forecast all significant eruptive episodes on Mount St. Helens several hours to weeks in advance. Data for forecasts comes principally from monitoring of seismic activity, ground deformation, and volcanic gas emissions. When a volcanic eruption is forecast, scientists and managers work closely with local officials to issue an advisory, alert, or update. Data from the volcano are also being compared with those collected in Hawaii. The volcanoes within the two regions exhibit widely different eruptive styles—Hawaiian shield volcanoes tend to erupt nonexplosively and exude highly fluid liquid lava, whereas Cascadian composite volcanoes tend to erupt explosively and eject highly viscous lava and pyroclastic ash and debris. But even our growing understanding of the signs and hazards of an impending volcanic eruption are fruitless without adequate emergency planning and implementation—a problem all too well demonstrated during the 1985 eruption of Nevado Del Ruiz.

Colombia, 1985: Nevado Del Ruiz

On the night of November 13, 1985, a sudden explosion rocked the crater of Nevado Del Ruiz. The blast triggered a pyroclastic flow of hot gas and debris that surged from the volcano's summit. At the volcano's peak there were extensive fields of snow and ice. Almost instantly, they were melted by the pyroclastic flows and gushed off the volcano in huge torrents of hot water, sediment, and debris. As water poured off the mountain, more rock and sediment were incorporated into the growing mud flow. Evidence suggests that some of the flows may have been nearly 46 meters deep.

The small town of Armero sat about 48 kilometers from the volcano's crater. Less than three hours after the initial blast, a river of mud, boulders, and trees rushed through the town, burying 20,000 people. It was a horrific tragedy, and one that could have been prevented.

The eruption of Nevado Del Ruiz in 1985 was smaller and less violent than the 1980 Mount St. Helens blast, yet many more people perished. A combination of natural and human factors explains the tragedy and has proved a costly lesson. Signs of an impending eruption may have actually begun in 1984 when Nevado Del Ruiz began to tremble and spew ash. In response, Colombian geologists began to examine the volcano's eruptive history by mapping the sediments of past eruptions. What they found was not good. Previous eruptions had created powerful mud flows that had rushed down the volcano's sides through stream valleys, inundating nearby towns. In 1845, a mud flow had killed about 1,000 people in the town of Ambalema, 32 kilometers to the southeast. A map was made outlining the volcanic hazards posed by Nevado Del Ruiz. Tragically, it was completed only days before the eruption, and when it was given to local authorities, its dangerous implications went unheeded. When the volcano erupted, the local townspeople were unaware of the eruption or the danger it posed. A flood-warning system placed above the town, similar to those used at the time in other areas of the world, could have provided enough time for people to respond. And with the appropriate education, they would have known to quickly move to higher ground, out of the path of the deadly mud flows.

When volcanoes erupt, the dangers are both numerous and lethal. Powerful explosions, searing pyroclastic flows, massive avalanches, falling ash, and raging mud flows all are possible. In rare instances, slower-moving lava can also pose a serious threat. The tragedy of Nevado Del Ruiz galvanized scientists and emergency management personnel to develop new strategies to help reduce the danger to those living near volcanoes. In 1991, when Mount Pinatubo erupted, these new plans were put into action and saved thousands of lives.

Philippine Islands, 1991: Mount Pinatubo

For five hundred years on the small Philippine island of Luzon, Mount Pinatubo quietly slumbered. Many in the region did not even realize that the nearby slopes and jagged peak were part of a sleeping volcano. Rather, it was a relatively peaceful mountain whose heavily vegetated sides and flat plains provided fertile soil and good living. The nearly level and flat ground fronting Mount Pinatubo were also well suited for the giant runways of a large U.S. military base, and so Clark Air Base and Subic Bay Naval Station were built on the site. In total, up until 1991, nearly 1,000,000 people lived in countryside towns, hillside villages, and military bases at the foot of Mount Pinatubo.

The first indication that Mount Pinatubo's quiet dormancy was soon to end may have come as early as July 1990. A large earthquake, magnitude 7.8, struck just 100 kilometers northeast of the volcano's summit. However, following a few landslides, small quakes, and bursts of steam, Pinatubo returned to its relatively calm slumber. Then in March and April 1991, the mountain came wide awake. Rising magma within the volcano triggered numerous small earthquakes and caused violent bursts of steam that created three large craters on the volcano's northern flank. Throughout the next two and a half months, thousands of small tremors shook the region, and huge quantities of sulfur dioxide gas were released into the air.

By early June 1991, a dome of swelling magma and oozing lava had grown on Mount Pinatubo's summit. Then, on June 12, the volcano's first true eruption occurred, and it was spectacular. Gas-charged magma breached the surface and exploded high into the sky, fueling a burgeoning mushroom cloud of gas and ash. On June 15, fed by more gas-rich magma from below, the volcano unleashed one of the largest, most explosive eruptions of the century (*Figure 3.10*). Ash and gas skyrocketed upward, producing an enormous cloud 35 kilometers high and 400 kilometers wide. Fine ash was driven high into the atmosphere, picked up by upper-level winds, and spread around the world. The sky in the nearby villages and towns turned dark, and volcanic cinders and ash rained down. To make matters worse, Typhoon Yunya arrived on the scene at about the same time. Hurricane-strength winds swirled in all directions, and heavy rain turned falling ash into a heavy fluid, something like wet concrete. Searing pyroclastic flows and huge avalanches of mud and debris mixed with the storm's torrential downpours and surged off the volcano. Streams and rivers were filled with torrents of steaming mud.

Afterward, in the once-deep valleys surrounding Mount Pinatubo, deposits of volcanic debris rose to nearly 200 meters. At the volcano's summit, removal of magma and rock during the cataclysmic blast had created a 2.5-kilometer-wide caldera.

Despite the enormity and devastation wrought by the eruption of Mount Pinatubo and the relatively dense population in the region, only 300 people died. Most fatalities resulted from the collapse of roofs under the weight of rain-soaked ash. Estimates suggest that at least 5,000 people and hundreds of millions of dollars in property and equipment were saved through the cooperative efforts of scientists and local officials.

In April 1991, when Mount Pinatubo began exhibiting signs of renewed activity—numerous small earthquakes and steam blasts—scientists from the Philippine Institute of Volcanology and Seismology stepped up efforts to monitor the volcano. A danger zone around the mountain 10 kilometers in

Figure 3.10
The eruption of Mount Pinatubo, June 1991.
Courtesy U.S. Geological Survey.

radius was quickly established. Soon scientists from the U.S. Geological Survey joined the effort to monitor the volcano and assess the hazards and likelihood of an eruption. The joint Philippine-American team rapidly went to work and set up an extensive network of modern monitoring instruments around the volcano, and examined volcanic deposits throughout the region to determine its eruptive history. Once again, the news was not good. Geologic data suggested that during previous eruptions, huge mud flows had cascaded off the volcano and thick deposits of ash indicated that an explosive type of eruption was likely. And measurements of seismic activity, deformation, and gas emissions suggested that a major eruption was immanent. Facing an impending eruption with potentially disastrous consequences, scientists warned the local authorities. To volcanologists, the threat was obvious, but in some cases they needed a means to convince others of the severity and seriousness of the situation. Reportedly, one of the most influential tools was a video entitled "Understanding Volcanic Hazards" produced by well-known volcanologists Maurice and Katia Kraft for the International Association of Volcanology. The video so vividly shows examples of pyroclastic flows, mud flows, and ash fall that it convinced many people of the risks involved and the need to evacuate. Local officials successfully ordered and implemented an evacuation. More than 75,000 people safely left the area, and both personnel and equipment were evacuated from nearby military bases. In addition, warnings were issued to commercial aircraft about the hazards posed by the towering ash cloud. Unfortunately, some jets flying far to the west of the Philippines sustained damage from ash circulating in the atmosphere. It is estimated that nearly 20 million tons of sulfur dioxide were released into the upper atmosphere during the 1991 eruption. The global dispersal of gas and ash from Pinatubo lasted for months and is thought to have caused a worldwide drop in temperatures of about 0.5° C.

A staggering amount of volcanic debris flowed and fell from Pinatubo, burying hundreds of square kilometers under thick deposits of mud and ash. To add insult to injury, in the years following the 1991 eruption, volcanic mud flows continued to periodically wreak havoc on the countryside. In some areas, ash and debris from the main eruption dammed up rivers and streams, creating new temporary lakes. Over time, the lakes eroded or overtopped their constraints and triggered new and powerful mud flows. Because these mud flows often occur in the absence of a heavy rain, they rage with no warning. Dozens of people have been killed and some hundred thousand people left homeless as torrents of mud have periodically flowed off the volcano's slopes. New towns have been built on higher ground out of the path of dangerous mud flows, but many remain in areas at risk.

The eruption of Mount Pinatubo in 1991 could easily have had the devastating impact and tremendous loss of life as the eruption of Vesuvius in 79 A.D. or Mount Pelee in 1902. All were huge explosive eruptions from composite volcanoes that spewed burning gas, ash, and raging flows of volcanic debris. Unquestionably, the difference in the deadliness of these events stems from our modern understanding of volcanoes, our ability to monitor and assess volcanic hazards, and the planning and implementation of evacuations. Efforts prior to and during the eruption of Mount Pinatubo in 1991 dramatically illustrate the very best of international cooperation and the importance of science to society. There are few equivalent examples in which scientific knowledge put to work helps so dramatically to save lives and property.

Forecasting Volcanic Eruptions

Will we ever be able to predict volcanic eruptions better than was done during the 1991 eruption of Mount Pinatubo? Scientists did not forecast the exact time and size of the eruption, but they accurately predicted that a blast was imminent and successfully prompted a life-saving evacuation. This may be as much as we can expect to do when it comes to forecasting volcanic eruptions. Like the prediction of weather and earthquakes, forecasting the eruption of a volcano is probabilistic in nature. Scientists can, with the appropriate information, assess the likelihood of an eruption occurring within a given time frame, but cannot pinpoint just when it will happen. In general, we know where volcanic activity is most likely to occur—along the edges of the tectonic plates, particularly behind subduction zones, and at hot spots. And we know that, just as with earthquakes, there tend to be more small eruptions than large ones. Mount Pinatubo illustrates that for individual volcanoes, given certain information, scientists can predict the likelihood of an eruption and what form it might take. To make these types of forecasts at least two types of data are needed, information on the volcano's day-to-day behavior and its eruptive history.

In an ideal world, for each volcano on Earth, scientists would have a long-term record of its eruptions and how it has behaved before, during, and following an event. But alas, reality sets in. For most volcanoes there are few records of pre-sent-day activity or past eruptions, and what records are avail-able span only a short time relative to the life of the volcano. In the absence of a large number of records stretching over a long duration, the probability of an eruption must be determined based on changes in a volcano's current activity. Monitoring can provide a measure of this change. Regular measurements of earthquake activity, deformation, and gas emissions can be compiled into a long-term "diary" for an individual volcano. Incoming data can then be carefully examined to look for changes from the "normal" baseline that might indicate the movement of magma or gas preceding an eruption. It is an expensive task that requires lots of instrumentation and trained, experienced personnel. In many regions of the world, particularly in poorly developed countries, volcano monitoring may be impossible or may take a back seat to higher-priority problems like hunger and disease.

Today, volcanologists concentrate their research efforts on intensely monitoring and studying specific volcanoes to learn how they erupt (past eruptions are indicated by geologic deposits) and what the warning signs are. Volcano observato-ries provide sites to closely monitor and intensely study vary-ing types of volcanoes. But there are many potentially dangerous volcanoes, and only a few that are regularly moni-tored or studied in depth.

With this in mind, and in the aftermath of the disastrous eruption of Nevado Del Ruiz, the U.S. Agency for International Development and U.S. Geological Survey established the Volcano Disaster Assistance Program. The program was designed to provide rapid response and assistance to regions faced with a threatening volcano. As part of the Volcano Disaster Assistance Program, a group of scientists stand by prepared to mobilize, with portable volcano-monitoring equipment, should a crisis develop anywhere in the world. A country faced with a restless volcano must, however, make an official request to the U.S. Department of State before a response team can be dispatched. On site, volcano experts work with local scientists to set up monitoring instruments and provide timely information and analysis to help local officials manage the situation and make the appropriate decisions. The success of the program was dramatically illustrated at Pinatubo in 1991, and since then it has provided assistance during situations in Central and South America, the Caribbean, Africa, Asia, and the South Pacific. Response teams were also recently dispatched to Mexico and the Caribbean Island of Montserrat.

In addition to forecasting the likelihood of a volcanic eruption, it is also important to provide information on the type of eruption that is likely to take place. If the volcano has erupted in recent history, then this information may be readily available. In most cases, little is known about a volcano's eruptive history, and scientists must rely on information contained within the sediments surrounding the volcano. Team efforts are quickly made to map deposits laid down during previous events.

Scientists may never be able to say exactly when and how a volcano will erupt, but even in its short history, volcanology is providing the know-how and means to forecast the probability of an eruption and what it will likely entail. Over the last decade scientists have had great success in forecasting eruptions at heavily monitored sites like Hawaii and Mount St.

Helens, and in regions that are instrumented once they become active, such as Mount Pinatubo and Montserrat.

Volcanic Hazards and Risks

A hazard is defined as a source of danger, whereas risk means the possibility of loss or injury. Although these two terms sound similar, they are actually quite different. When it comes to volcanic hazards and risks, the difference is critical, because while we can reduce risks from volcanic eruptions, rarely can we minimize the hazard. In other words, we cannot stop a volcano from erupting, but we can do things to minimize and reduce the possibility of injury or property loss during an eruption.

The first essential step in minimizing the risks from a potentially active volcano is to define the hazard (*Figure 3.11*). Typically, scientists create a map of a region's volcanic hazards based on all the collected and available data. The hazards map outlines where and what kind of dangers exist should a nearby volcano erupt or activity intensify. For instance, large blocks and bombs ejected out of a volcano during explosive eruptions tend to fall within about 3 kilometers of a volcano's vent—thus large rock falls are a volcanic hazard for all areas within a 3-kilometer radius of a potentially active, explosive volcano. Smaller cinders may pose a hazard to a wider area, and ash can threaten regions far distant from an eruption site. Ash fall poses a particularly dangerous threat to communities downwind of a volcano and commercial jets flying in the region. At least eighty commercial aircraft have been damaged while inadvertently flying into or through an ash cloud over the last

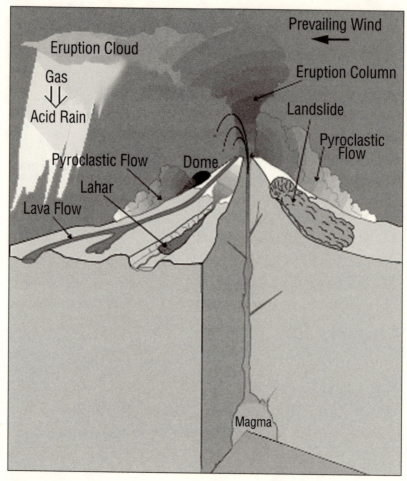

Figure 3.11
Volcanic hazards. Courtesy U.S. Geological Survey.

fifteen years. The emission of large quantities of noxious gases, such as sulfur dioxide, carbon dioxide, and fluorine, also poses a threat to regions around a volcano, especially those downwind. Thus knowing the general pattern of winds over a volcano is also important when creating a volcanic hazards map.

In Hawaii, current, long-lived eruptions on Kilauea have created a new volcanic hazard, dubbed *vog*. Vog, or volcanically derived smog, is created when volcanic gas emissions combine and interact with oxygen, moisture, dust, and sunlight in the atmosphere to form a persistent haze of unhealthy air. Vog can cause respiratory problems, kill vegetation, and contaminate drinking water.

A volcanic hazards map is also used to show where erupting lava is most likely to flow. The potential paths of pyroclastic surges, debris avalanches, and mud flows must also be delineated. Mud flows are particularly likely to occur where volcanoes are topped with snow and ice. Because earthquakes generally accompany volcanic eruptions, they too are considered a volcanic hazard. And because both quakes and avalanches can trigger tsunamis, tsunamis are volcanic hazards as well. In retrospect, these hazards seem obvious, but consider that many volcanoes lie dormant for centuries or even millennia. In the countryside surrounding a dormant volcano, people may be completely unaware of the potential dangers. In 1986, another relatively rare, though no less deadly, volcanic hazard made itself known to the world.

On August 26, 1986, 1,700 people were killed when an enormous cloud of carbon dioxide was released from Lake Nyos, Cameroon (*color plate 14*). It is unclear what triggered the deadly event, but the underlying cause has since been determined. Lake Nyos is a volcanic lake; it sits over the active vent of a volcano. Deep beneath the lake, magma rich in carbon dioxide comes into contact with groundwater. As a result, groundwater seeping into the lake becomes charged with a high concentration of carbon dioxide. Over time, this carbon dioxide accumulates in the lake's deep waters. Here, a layer of dense cold bottom water sits trapped beneath an upper layer of less dense water, warmed by the sun. Because the lake is deep and sheltered from the wind, the layers normally remain stratified and don't mix. Any event or disturbance that causes the lake to overturn or the deep gas-rich water to rise can trigger a major release of highly concentrated and lethal carbon dioxide gas.

Research following the deadly 1986 event has also revealed that the gas content on the bottom of the lake is still very high and increasing. The probability of another lethal gas release is therefore also rising. However, now that the hazard and its underlying cause are known, efforts are being made to reduce both the hazard itself and the risk. A plan has been proposed to pump the gas-rich bottom water out of the lake and prevent it from reaching dangerous levels. Plans are also being made to educate local communities about the risks posed by volcanic lakes.

Reducing Risk

Minimizing the risks posed by volcanic hazards entails lowering the probability that people will be injured or property lost. The most obvious way to reduce risk is to move all those residing in regions of hazard out of harm's way. However, the best solution is not always possible or realistic, particularly because many volcanoes lie dormant for so long. Other ways to reduce risk include the establishment of monitoring networks, the development of warning and evacuation plans, and education and preparation within local communities. In some rare instances, protective structures have been built or the course of lava flows altered.

Volcano monitoring systems can clearly detect indications of change in a volcano that often precede an eruption. A volcano may exhibit clues indicative of unrest months to years before an eruption, and scientists are getting better and better at measuring and recognizing the warning signs. New techniques are also under development to detect and warn of real-time volcanic threats. A volcanologist with the U.S. Geological Survey recently developed a highly durable and inexpensive sensor to detect volcanic mud flows. A system of specially designed monitoring stations is installed downstream of a volcano; each station is equipped with a high-frequency seismometer that senses ground vibrations characteristic of a mud flow. The instrument is specifically designed to distinguish the ground motion of a mud flow from that of volcanic or earthquake activity. New mud flow detection systems have been installed on several volcanoes in the United States, Indonesia, Philippines, Ecuador, Mexico, and Japan.

Scientists are also currently working to develop less expensive but relatively accurate GPS-based deformation monitoring devices. Although each individual GPS receiver may be less accurate than the more expensive models, a large number deployed around an individual volcano can together give an accurate measurement of relative ground movement. In addition, if one is destroyed or stops working, it is relatively easy to replace. These types of receivers have already been deployed on Popocatepetl in Mexico, Mauna Loa in Hawaii, Taal in the Philippines, and in the Long Valley Area of California. Software is currently under development that will transfer and analyze the GPS data and present it in a useful form for hazard assessment.

Another real-time hazard monitoring and warning system has been devised for the threat to commercial aircraft from ash clouds. Pilots are often unable to distinguish ash clouds from ordinary clouds, and winds may carry volcanic ash far from the erupting volcano. In 1989, Redoubt Volcano in Alaska had been erupting for ten hours when Flight 867, inbound for Anchorage, encountered its ash cloud (*color plate 15*). The cockpit of the plane became filled with foul sulfur-smelling smoke (rotten eggs), and suddenly the engines quit. For five very long minutes the crew struggled to restart the engines and calm the terrified passengers as the jet dropped nearly 4,000 meters. Luckily, the expert crew was able to restart all the engines and land safely in Anchorage. The damage to the plane was estimated at $80 million, which included the replacement of all four engines. Other incidents in which commercial jets temporarily lost power and incurred extensive damage occurred over Indonesia in 1982 and the Philippines in 1991. Within the North Pacific, nearly 100 dangerously active volcanoes lie under heavily trafficked air routes. Scientists and government agencies are now working with the aviation community to issue timely and informative advisories on volcanic activity and ash clouds. Funding has even been provided by the U.S. Federal Aviation Administration to triple the number of volcanoes monitored by the Alaskan Volcano Observatory. Advisory notices provide the geographic coordinates of active ash clouds, their trajectory, a map of the cloud shape and size, and continuing updates. Much of the information in the warnings comes from satellite images provided by the National Oceanic and Atmospheric Administration.

Clearly, a vital part of any warning scheme—along with a foundation of good science and detection—is excellent communication. Science-based forecasts can only be effective if they are adequately explained to the appropriate decision-making authorities. Adequate and meaningful information must also be relayed from local or regional officials to the public and local emergency groups. These efforts must communicate the hazards posed by an impending eruption as well as the risks involved. The use of the volcano video during the Mount Pinatubo eruption was an excellent and convincing form of communication. In some regions where potentially dangerous volcanoes loom, a system of color-coded messages has been established to warn and inform the local authorities and the public about volcanic unrest and potential eruptions.

In California, the state's Office of Emergency Services and local authorities, in collaboration with the U.S. Geological Survey, have established a response system to volcanic activity in the Long Valley Area of eastern California. A condition green indicates that no immediate risk is posed by volcanic activity, and routine monitoring continues. An upgrade to a yellow volcano watch means that intense volcanic unrest has been detected. For instance, monitoring instruments may have detected an earthquake swarm containing at least one magnitude 5 quake or deformation on a volcano's summit. If a yellow volcano watch is issued, monitoring efforts are intensified, an emergency field headquarters is established, and the appropriate authorities are notified. An orange volcano warning is issued when an eruption is likely to occur within hours or days as evidenced by shallow magma movement. A geologic hazard warning is issued to the regional governors, who then inform the public. Intensive monitoring on site is continued. And when a red volcano alert is issued, it means an eruption is underway. Monitoring and communications continue, and the authorities are continually updated on the eruption's progress and probable events. Estimates are that in the Long Valley Area, a red volcano alert could occur about once every few hundred years.

Similar color-coded warning systems exist in other regions, such as Alaska, the Pacific Northwest, and Hawaii. Because the volcanoes in each of these regions exhibit different types of activity and pose differing hazards, the warning schemes vary slightly. Emergency service communication lines and radio and television broadcasts are used to relay warning messages. The Internet has become a rapid means of providing comprehensive information not only on specific warnings' but also on volcanic activity worldwide (see the listing at the end of the book). A vast amount of information is now readily available about volcanoes, where they are active, past eruptions, hazards, and new research findings. But even where warning systems are in place (or can be put in place at short notice) and information available, appropriate response plans and education are needed to reduce volcanic risks.

Local authorities in areas of potentially active volcanoes must design response plans to implement and enforce timely evacuations. As part of this, the public must be educated on how to respond to evacuation orders or warnings. Volcanologists work with groups throughout the world to educate people about volcanic hazards and the appropriate response to warnings. If the community of Armero, Colombia, had known to move to high ground during the eruption of Nevado Del Ruiz, thousands of lives could have been saved. Most did not even know that volcanoes can trigger deadly mud flows, let alone that past flows had swept through their valley.

Proactive education about volcanic hazards and risk are especially important in regions where, for the last 100 years or more, nearby volcanoes have been at rest. No memory of the dangers posed by a volcano may exist within the region. The snow-topped peaks of many a majestic mountain may not be as serene and peaceful as they appear. Even in the United States, many are unaware of the volcanoes that loom nearby and pose a serious threat, possibly in the not-so-distant future.

Regions at Risk

Volcanic activity occurs most commonly around the Pacific Rim, in the infamous region known as the Ring of Fire. Worldwide, there are more than 1,500 potentially active volcanoes. Recent volcanic activity has occurred or is continuing at volcanoes in Guatemala, El Salvador, Colombia, Ecuador, Mexico, Montserrat, Italy, Tanzania, Indonesia, Vanuatu, New Zealand, Alaska, Hawaii, and Kamchatka. Within the continental United States, volcanoes in the Long Valley Area of California and the Pacific Northwest are of specific concern.

One week after the 1980 Mount St. Helens eruption, another potential site of volcanic activity, Long Valley in eastern California, awoke and caught the attention of many people. Volcanic deposits in the region suggest that if Long Valley were to erupt, the blast could be 500 times greater than that of Mount St. Helens.

Some 760,000 years ago, a colossal volcanic eruption occurred that covered east central California with burning flows of hot ash and debris. Airborne ash fell as far east as Nebraska, and the removal of magma from beneath the volcano resulted in the creation of a 16-kilometer-wide, 32-kilometer-long caldera. Today Long Valley, California, sits atop the eastern half of the caldera and its underlying smoldering magma (*Figure 3.12*). Natural hot springs in the region are evidence of the hot molten material that lies below. Following the earthquake activity in 1980, scientists began taking a closer look at the potential for an eruption in the Long Valley region. Mammoth Mountain and other nearby areas are popular sites for recreational activities such as skiing, hiking, and camping.

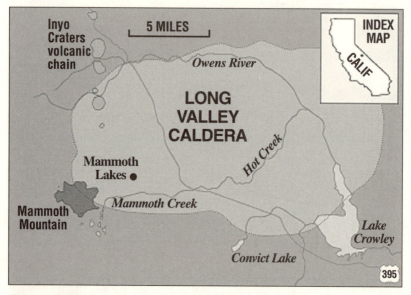

Figure 3.12
Long Valley area and caldera, California.
Courtesy U.S. Geological Survey.

Scientists soon discovered that since 1979, a dome had been slowly growing in the caldera. During the 1990s, scientists found areas of dead or dying trees on Mammoth Mountain. Studies revealed that carbon dioxide seeping up through the ground and accumulating in the soil was killing the trees. Fractures created during earthquakes in 1989 are thought to have released the gas from the underlying magma into the soil. Nearly 100 acres of trees have been destroyed by subterranean carbon dioxide emissions. No definitive evidence of current magma movement has been detected, but earthquakes, deformation, and gas emissions continue to be carefully monitored in Long Valley, and an alert system is in place.

In the last few thousand years, more than 100 volcanic eruptions, mainly explosive in nature, have occurred in the Cascade Mountain Range of the U.S. Pacific Northwest. The Cascade Range includes a chain of at least thirteen potentially active volcanoes, stretching from northern California to British Columbia. Mount St. Helens lies on this linear belt of sleeping giants. Because the volcanoes in the Cascades can remain dormant for many years, the hazards and risks they pose often go unrecognized. These are composite volcanoes that lie behind and on top of the Cascadian Subduction Zone, and when they erupt, they tend to be explosive. Searing pyroclastic and lava flows have surged off these mountains in the past, as have debris avalanches and raging mud flows. One has only to look at the snow-topped crown of Mount Rainier to imagine the mud flow that would be created should this slumbering volcano awaken (refer to *Figure 3.5*). Scientists continue to closely monitor the volcanoes of the Cascade Range, study their present-day activity and eruptive history, and educate the local populations about the hazards and appropriate response should a warning or alert be issued.

As the Earth's population continues to grow, more and more people will be living and recreating in the shadow of volcanoes. Deposits of volcanic ash make for especially fertile soil, volcanic peaks provide majestic scenery and sites for recreational activities, and hot springs are natural wonders that many enjoy. While volcanology is a relatively young science, it is one that is rapidly growing both in knowledge and recognition. Modern technology is providing the means to better study volcanoes as the recognition of their dangers provides the impetus. Safer and more accurate monitoring devices are being developed, and better cooperation among experts and those in need of assistance is being fostered. Unquestionably, intensified efforts to better understand the science and nature of volcanoes and to more accurately forecast eruptions will pay off in terms of saved lives and property.

Chapter 4

Tsunamis

Papua New Guinea, 1998

The evening of July 17, 1998, started just like many others along the tropical shores of northeastern Papua New Guinea. As the day came to an end, people within several of the fishing villages on the narrow, sandy strip of land adjacent to Sissano Lagoon prepared their evening meals. It was the start of a four-day national holiday, and many had gathered around the lagoon to enjoy the evening. Just before 7:30 P.M. a loud bang that reportedly sounded much like a jet engine echoed throughout the region. The earth began to shake. Few took notice, however; earthquakes are fairly common in the area, and this one seemed relatively tame. Just minutes after the loud noise, the sea level began to fall, moments later a towering wall of water, some 10 meters high, came rushing toward shore. In all, three waves of gigantic proportion, tsunamis 7 to 15 meters tall, struck the Papua New Guinea coast that night, sweeping over the narrow stretch of sand fronting Sissano Lagoon, home to 12,000 people (*color plate 16*). Hardest hit were the fishing villages of Warupu, Arop, and Malol. Nearly all the village homes were washed away, and some 3,000 people were killed or lost. Twelve hours went by before the disaster was even discovered by the outside world, when a helicopter carrying people to a nearby lumber mill flew over the area. Initially, an offshore earthquake was thought to have created the monster waves. However, as the seismic and damage reports flooded in, scientists began to question how a relatively moderate-sized earthquake, with a moment magnitude of 7.1, could produce such a mammoth and devastating tsunami. In fact, it is not at all clear why some earthquakes produce tsunamis, while others that are expected to, do not. To learn more about the specific nature of the Papua New Guinea tsunami, the International Tsunami Survey Team was assembled and dispatched to the disaster site.

In the past six years there have been ten major tsunami events (Nicaragua, 1992; Flores, 1992; Okushiri, 1993; East Java, 1994; Mindoro, 1994; Kuril Islands, Russia, 1994; Manzanillo, 1995; Irian Jaya, Indonesia, 1996; Peru, 1996; and Papua New Guinea, 1998). In response to each, an international team of tsunami experts has been sent to the scene to gather information and learn as much as possible in the tsunami's aftermath. The purpose of the team is to collect data that will lead to a better understanding of tsunamis and ultimately save lives through improved hazard and risk assessment, education, prediction, and warnings. Much of what we know today about tsunamis comes directly from observations or eyewitness accounts immediately following an event. But scientists must act quickly, or valuable information may be lost because of cleanup activities and subsequent storms.

At each site, the International Survey Team works closely with local officials and disaster relief efforts to survey the aftermath of the tsunami, taking particular care to document evidence of how high, far, and fast the water flowed. Marks on walls and windows left by dirty flowing water are used to document the run-up height, a measure of how high the water reached. The run-up distance, the extent of the tsunami's penetration inland, is judged based on evidence such as the boundary of debris and where vegetation has been killed by saltwater. Every mark used as a basis for run-up estimates is photographed and precisely located using the Global Positioning System (GPS). Eyewitness accounts also provide invaluable information on the characteristics of a tsunami and the details of the wave right before it struck. Scientists compare the observations made by the team with computer simulations and attempt to model the height and speed of the tsunami as it approached and struck land, as well as the distance it traveled inland. In practical terms, experts hope that modeling can help tell people how far and how high they might need to run to escape an approaching tsunami in future events. The data collected by the International Tsunami Team in Papua New Guinea and elsewhere is beginning to reveal much about the science and nature of tsunamis.

Scientists now suspect that the 1998 Papua New Guinea tsunami was not created by the earthquake itself, but was triggered by a massive underwater landslide caused by the quake. In some tsunami events, the effects are felt over great distances along one or more coastlines. In Papua New Guinea, the brunt of the tsunami's impact fell along a relatively limited 20-kilometer section of the coast. Those living within the fishing villages along the narrow strip of land fronting Sissano Lagoon were particularly vulnerable. The communities around the lagoon had been built on a low-lying piece of land with the ocean at its front and a lagoon to the back, so there was no escaping the onrush of water.

The Papua New Guinea (PNG) event also dramatically illustrated that powerful tsunamis can be triggered locally; they do not have to come from distant sites to have disastrous results. For many years, tsunami experts focused their attention on tsunamis that are created by large earthquakes, cross the ocean, and then crash down upon distant shores. Both local and so-called distant or remote tsunamis can be catastrophic, but gigantic waves generated far from shore take time to travel across the ocean and may allow for an hour or more of warning. Locally triggered events provide little more than a few minutes of warning. Researchers continue to study and model the PNG event in an effort to better understand tsunamis, and in particular those that are triggered locally.

While we understand more than ever before about what causes tsunamis and how they travel across the open ocean and behave at the coast, there remain many unanswered questions about these towering walls of water. An old but still puzzling example comes from a catastrophic event that took place more than a century ago in Indonesia. On August 27, 1883, the volcano on the island of Krakatau unexpectedly erupted in a violent and powerful blast. About 40 kilometers to the east of Krakatau lay Indonesia's Sunda Straits and the shores of Java and Sumatra. When the volcano blew, it sent steam and ash high into the stratosphere and triggered waves that grew to more than 30 meters in height and raced across the sea at 1,130 kilometers per hour (700 mph). Thirty-five minutes after hearing the explosion, people on the west coast of Java reportedly saw huge mountains of water rolling toward shore. When the tsunamis hit, palm trees snapped like matchsticks and houses were crushed and washed away. A gunboat, the *Berouw*, was lifted up, carried two miles inland, and dropped abruptly. In all, 165 settlements were destroyed and more than 32,000 people killed. Scientists continue to debate how Krakatau's eruption actually generated a tsunami.

Three explanations have been offered for Krakatau's catastrophic waves. One theory suggests that the mixing of seawater and gas-rich magma within the volcano unleashed a powerful explosion beneath the sea. Alternatively, some believe that the tsunami was triggered after the eruption, when a tremendous surge of seawater was drawn into the volcano's empty magma chamber. A third hypothesis suggests that a massive flow of volcanic debris cascaded into the sea and created the monster waves. Researchers have run computer simulations of the event using each of the three proposed mechanisms of tsunami formation.

When these simulations are compared to evidence recorded at the scene and historical accounts, the underwater explosion seems to fit best. However, in extensive surveys throughout the region, other scientists have found large submarine deposits of ash and pumice. These researchers believe that the true cause of the tsunami was the pyroclastic flows, not an undersea explosion. In any case, it remains unclear how Krakatau's eruption physically created such huge walls of water. In other instances, tsunamis have been known to inexplicably fade away to little more than a high tide before striking the coast. What causes such extreme differences and controls the shape of a tsunami when it hits the shore? How much have we really learned over the past century of research, and will we ever be able to accurately predict and warn of the sea's impending wrath?

Tsunami or Tidal Wave?

From the Japanese, the term *tsunami*, meaning "great harbor wave," refers to a seismic sea wave, one that is triggered by a seismic event. Why do the Japanese call tsunamis harbor waves? In Japan, historical documentation of tsunamis goes back 1,000 years and suggests that they attack Japan's shores on average about once every decade. When tsunamis enter a harbor, they are amplified by bouncing off of the harbor's embankments and combining. Alaska's 1964 Good Friday earthquake triggered monstrous tsunamis that swept into harbors throughout the region, including at Anchorage, Valdez, and Seward, and caused catastrophic destruction. Not only can these waves reach towering heights in a harbor, but often the water motion persists for many hours. Most likely, in early Japan, harbors were where most people witnessed and recognized these giant waves as something out of the ordinary—thus seismic sea waves became known as great harbor waves. In Spanish, the word for tsunamis is *maremotto*, meaning a trembling sea.

Tsunamis were also well known to the ancient Greeks; the eruption of a volcano on Thera, around 1500 B.C., triggered a tsunami that destroyed the Minoan civilization on the island of Crete, 128 kilometers to the south. Modern estimates put the height of the huge wave at close to 30 meters (100 feet).

The Greek philosopher Plato mentions in his book *Timaeus* that the continent of Atlantis was lost to "tidal waves." Many now believe that he took poetic liberty with the destruction of Thera and invented Atlantis, which later lent its name to the Atlantic Ocean. Although *tidal wave* is an exact translation of the Greek word for tsunami, most believe that the ancient Greeks realized the difference between the tides and tsunamis. One can only speculate why the term *tidal wave* has fallen into such common use, or more precisely misuse, in the last twenty years or so. Perhaps the term *tsunami* is too foreign and not as visually descriptive as *tidal wave*. Or maybe it is the mistaken belief that tsunamis always resemble walls of water moving up the coastline, somewhat like a tide. In reality, tsunamis exhibit little that is similar to the gradual ebb and flood of the tide (or a true tidal wave or tidal bore), so the term *tidal wave* is not only incorrect with regard to its origin, but also inappropriate descriptively.

Tsunamis vs. Wind Waves

A tsunami can be created by the sudden movement or disturbance of the sea floor, a submarine explosion, or the impact of a large object in the sea, such as a landslide or an asteroid. These geophysical events trigger a series of fast-moving, low, long waves that radiate outward in all directions. In contrast, most of the ocean's waves are generated by wind dragging or disturbing the surface of the sea. When you blow over the surface of the liquid in a teacup, small wind waves are created. If you shake the teacup, larger waves slosh back and forth. Real-world sloshing waves are called seiches and are often observed

in lakes and reservoirs during or immediately after large earthquakes. Tsunamis are commonly generated in a similar fash-
ion, by the violent shaking and deformation of the sea floor.
Wind waves and seismic sea waves differ markedly as a direct
consequence of the differences in how they form.

Both wind waves and tsunamis are characterized by a
wavelength, the horizontal distance between crests or peaks; a
period, the time it takes successive peaks to pass a fixed point;
and height, the vertical distance from the wave trough to its
crest (*Figure 4.1*). Waves generated by the wind tend to have a
wavelength of between 0.02 and 130 meters and periods of
about 0.2 to 30 seconds. In contrast, tsunamis typically have a
wavelength hundreds of kilometers long and a period ranging
from ten minutes to over an hour. Wind waves vary in height
from tiny ripples on the sea surface to rare rogue waves 30
meters tall. Tsunamis, on the other hand, race across the open
ocean as a series of long, low-crested waves, usually less than

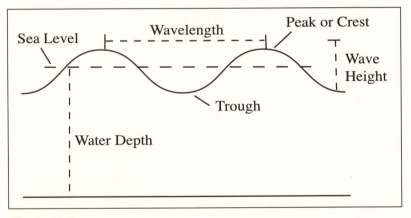

Figure 4.1
Wave terminology.

1 to 2 meters (1 to 6 feet) high. In fact, until they reach the coast, these mountains of water are benign beasts, virtually imperceptible to the human eye. A ship out at sea may sit completely unaware as a deadly tsunami passes beneath the hull.

In general, waves are considered deep-water waves if their wavelength is relatively small compared to the water depth through which they travel. They are little affected by the water depth or sea floor. In the open ocean, where depth averages about 4 kilometers (2.5 miles), most wind waves are deep-water waves—having a short wavelength relative to depth. In contrast, shallow-water waves are those with a long wavelength relative to depth. The depth and nature of the sea floor strongly influence how these waves propagate or travel. Because tsunamis have such long wavelengths, even when traveling through very deep water, they are considered shallow-water waves.

Although all waves appear to move water toward the coast, in reality the water just goes up and down, while energy is transferred forward. Waves transfer energy through the sea by creating an up and down or orbital motion of the water molecules. If you were to follow the motion of a parcel of water throughout the passage of a wave, it would go upward as the crest approaches, forward as it passes, downward into the trough, and backward as the trough goes by (*Figure 4.2*). This orbital motion of the water creates the wave and its illusion of forward movement. If the crest of a wave is higher than the trough is low (as in breaking), then the water itself does move slightly forward.

In wind-generated waves, the orbital motion of the water decreases with depth (with distance away from the wind). At a

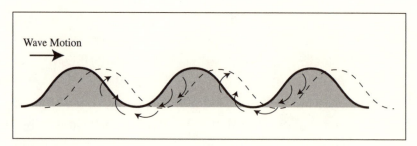

Figure 4.2
Orbital motion of a wave.

depth of about one-half the wavelength, the orbital velocity of the water is reduced by 96 percent. Consequently, because energy is transferred through the water's motion, in wind waves traveling through deep water, most of the energy is concentrated near the surface. Even for relatively large wind waves, the amount of energy being transferred is relatively limited to the near-surface waters. On the other hand, in a tsunami, the energy imparted to the water during its formation (e.g., an earthquake) sets the entire water column in motion (remember the teacup). Orbital velocities do not decrease significantly with depth, and although the wave height at the surface is relatively small, a few meters at most, the energy contained throughout the entire water column is huge. Furthermore, the rate at which waves lose energy is inversely proportional to their wavelength. So tsunamis not only contain lots of energy and move at high speeds, they can also travel great distances with little energy loss—a prescription for disaster for the coastlines they strike.

As waves travel through the sea, water moves up and down and energy is transferred toward the shore. The speed (e.g., in meters per second) of a wave can be estimated by dividing its wavelength (in meters) by its period (in seconds). So for deep-water waves, their speed is a function of their wavelength or period. For shallow-water waves, such as tsunamis, another factor comes into play—the water depth (remember a shallow-water wave is one whose wavelength is relatively long compared to its depth). In fact, the water depth becomes the governing factor; and, based on physics, the speed of a shallow-water wave can be estimated simply by multiplying the water depth by the gravitational force (acceleration of gravity, 9.8 m/s/s) and taking the square root. For instance, in typical ocean depths of 4,000 meters, a tsunami travels at a speed of nearly 200 meters per second or almost 700 kilometers per hour—the speed of a jet plane. When tsunamis enter shallower water they slow down; at a depth of 30 meters, a shallow-water wave travels at only 59 kilometers per hour (36 miles per hour). Because tsunamis travel as shallow-water waves, this simple calculation can be used to estimate their speed in the open ocean. If only predicting how tsunamis behave by the time they reach the shore were so simple!

Tsunamis are most often generated in relatively deep water, travel through shallow water, and then strike the shore. As they move toward the coast, tsunamis pass through varying depths and over complex sea floor topography. Changes in the depth and sea floor cause these long, low waves to continuously evolve and change shape. A tsunami generated from an earthquake off Peru may look entirely different along the Peruvian coastline compared to its appearance as it enters a bay in California, and different still when it strikes a beach in Hawaii. For scientists who study tsunamis or those involved with issuing warnings, these transformations present a real challenge. Even if it is possible to know the precise shape of a tsunami at its origin, and to estimate its speed, the entire evolution from generation to the target coastline must be predicted to provide meaningful evacuation warnings. Scientists employ computer models based on mathematical representations of wave motion to simulate and predict how tsunamis and other waves move through and evolve in the ocean. The foundation of all such models is Newton's law of motion, which relates the mass of a moving body and its acceleration to the force that drives the motion. Because the equations employed are very complex, their mathematical solutions are only approximations of the true values, for either long or short waves. In the coming years, ultra-efficient computations running on supercomputers will allow for much better wave modeling, particularly over long travel distances.

As a tsunami moves into shallow water, approaching the shore, it slows and steepens, sometimes breaking in a mountainous cascade of water. No longer is the simple shallow-water wave formula for speed applicable. And little is known about the speed of a tsunami when it breaks over dry land. Survivors indicate that tsunamis may travel as fast as 48 kilometers (30 miles) per hour over the land. Even the fastest sprinter cannot outrun a breaking, massive wall of water. The size and force of a tsunami striking the shore depends on the local topography, the shape of the shoreline, and the direction of approach. One of the highest documented tsunamis ever, occurred in 1971 in the Ryukyu Islands of Japan, where at the coast, a wave reportedly crested at a towering 85 meters (278 feet) above sea level.

When striking the coast, a wave's energy is transferred from the water to the land. In the case of a tsunami, it is this transfer that spells disaster. As waves enter shallow water or shoal, friction from the sea floor flattens the orbital motion of the water particles near the bottom. Three things then happen to the wave: its wavelength and speed decrease, and its height increases. Conceptually, imagine the bottom portion of the wave slowing, the top continuing at the same speed, and the entire wave "bunching up." When this happens, the wave steepens, the top overrides the bottom, and finally, it breaks. Essentially, the wave trips over its own feet. Because a tsunami has such a long wavelength, there is a lot of water to bunch up, and it can begin to slow and steepen far offshore, growing mammoth by the time it reaches land. As the wave plunges or spills onto the shore, its energy is released. The higher the wave, the greater the energy. In everyday waves this energy transfer is responsible for eroding beaches, moving sand along the shore, and creating rip currents. In tsunamis, the transfer of energy can snap trees, destroy villages, and carry boats far inland. Not all tsunamis break at the shoreline; some just swiftly submerge the shore and generate furious, swirling currents.

Many tsunamis make their entrance into the coastal zone as a leading depression wave (*color plate 19*). As the long wave in a tsunami begins to shoal, its "bunching up" draws in water from along the shore. Stories tell of people venturing out to collect fish that have been left stranded and flapping by a great retreat of the sea. But for those who partake of this all-too-easy harvest, the end comes quickly when, moments later, a tsunami crashes down. Obviously, a retreat of the sea should be a clue—run far, fast, and away, and if possible, up from the coast. But not all tsunamis provide even this warning to local populations. If the triggering event is far away or deep beneath the sea, there may be no indication of the wall of water soon to strike.

Okushiri Island, Japan, during a 1993 tsunami. This town too lay on the lee side of an island, yet it suffered a 20-meter high run-up from the tsunami.

Like other types of waves, tsunamis bend or refract according to variations in depth along a coastline. When a line of waves approaches the shore, the portion that first moves into shallow water begins to slow relative to the rest of the wave. The part of the wave still in relatively deep water continues to propagate at the same speed. The line of waves then appears to bend or curl toward the slowing portion of the wave. Standing on a cliff or flying low over the shoreline, you can see the bending of the waves (called wave refraction) around a cliff, pier, or shallow zone. Wave refraction can cause tsunamis to bend around islands, focus on headlands, or propagate into sheltered regions.

Tsunamis can cause sediment erosion or deposition, or they can tear apart coral reefs that lie in their path. Coastal regions that are low lying or are surrounded by steep cliffs or bodies of water are particularly vulnerable to tsunami damage. On June 3, 1994, an earthquake occurred within the Java Trench in the Indian Ocean. The magnitude 7.2 quake triggered a large tsunami that struck the coast of southeast Java and rolled on to hit southwest Bali. Two hundred people were killed, 400 injured, and 1,000 left homeless. Post-tsunami surveys found clues, such as trees with sand-encrusted bark and leaves, that indicated a run-up of 5 meters in west Bali and up to 14 meters in southeast Java. Many of the fatalities and injuries occurred because rivers had blocked evacuation routes. Several beaches were completely washed away, and eyewitnesses reported that on a 20-meter section of a nearby coral reef, about a meter's worth of surface growth had been literally shaved off. Later, a few large pieces of the reef were found on a nearby beach, and much of the eroded beach sediment was discovered offshore, forming broad sandbars. The same tsunami was documented along the northwestern Australian coast, where in one area a surge of water 3 to 4 meters high carried fish, crustaceans, and rocks nearly 300 meters inland.

The Trigger

While landslides, volcanoes, and asteroid impacts can all trigger tsunamis, by far the most common cause is submarine earthquakes. Even if the quakes themselves do not trigger tsunamis, associated landslides may do the trick. However, not all earthquakes generate tsunamis; in the past ten years, there have been more than 100 quakes of magnitude 6.0 or greater, yet only 12 of them reportedly created tsunamis. The pattern and extent of vertical ground deformation from an earthquake may determine why some quakes trigger tsunamis and others do not.

As described in the previous section on earthquakes, there are three types of faults: strike-slip, thrust, and normal (refer to *Figure 2.7*). Strike-slip or transform faults involve horizontal motion of the Earth's crust, while thrust and normal faults entail vertical motion. Submarine thrust and normal faults produce tsunamis because the sea floor lifts up or drops down, either pushing the water up or pulling it down—triggering wave motion. On the other hand, strike-slip motions do not generally cause vertical displacement on the sea floor, so typically there is little response in the overlying water. Most faults combine both strike-slip and thrust motions, but only those faults that have predominantly vertical displacement and create sufficiently large sea-floor deformations (or landslides) appear to trigger a tsunami.

In general, the larger the magnitude of an earthquake, the larger the area that is deformed. Scientists investigating how earthquakes trigger tsunamis measure the deformation area (the horizontal extent of displacement) and the deformation length (the amount of vertical slip or displacement on the sea floor). Earthquakes on thrust or normal faults create deformation areas that are substantially larger than those of strike-slip origin. Strong earthquakes not only deform larger areas, they do so by a greater amount.

In addition to an earthquake's magnitude, its depth also influences how much and how far the Earth's crust is deformed in a given event. The deeper the hypocenter or focus of an earthquake, the smaller the vertical deformation of the Earth's surface. Visualize bending thick and thin strips of rubber: if one applies the same force to each, the thicker piece will flex less than the thin. Actually, a deeper hypocenter allows the seismic energy to spread over a larger volume so that less energy reaches the Earth's surface. Earthquakes that originate below 30 kilometers rarely cause sufficient deformation to generate tsunamis. However, truly great earthquakes, like the 1960 Chilean event, can occasionally trigger tsunamis even if they occur at greater depths. Tsunamis have never been reported from an earthquake with a hypocenter deeper than 60 kilometers.

An earthquake with an epicenter that lies inland will create a tsunami only if it produces sufficient vertical deformation offshore on the sea floor. Therefore, only very strong inland thrust earthquakes, as compared to even moderate offshore quakes, are potential tsunami generators (unless, of course, they trigger a massive landslide into the sea). For example, the 1994 Northridge earthquake that violently shook Los Angeles resulted in vertical ground deformations of up to about 2 meters, but it did not produce a tsunami. Had the fault ruptured with the same ferocity about 64 kilometers (40 miles) to the west and offshore, it would have probably created a substantial tsunami inside Santa Monica Bay.

Until recently, earthquake-induced sea-floor deformation was believed to be the primary cause of most tsunamis. However, since 1992, a series of ten devastating tsunamis, including the 1998 Papua New Guinea event, have led scientists to question this long-held belief and to test the validity of computer models based on it. Would the models of earthquake deformation produce a tsunami that matched the wealth of new field observations? This type of comparison between field data and model predictions is referred to as validation or verification and is a crucial part of any scientific modeling effort. Without comparison to real-world data, we should question the predictive capability of any model.

Comparison of the modeled tsunamis with the field-collected data from the events in the 1990s revealed that sometimes even the best computer model fell short of predicting tsunami behavior. Scientists began to consider whether it was really a shortcoming of the models or if something other than earthquakes was generating the tsunamis—such as quake-induced landslides. Until then, tsunamis generated by mass movements were believed to be extremely rare, even though the highest vertical run-up ever recorded, 457 meters (1,500 feet), was the result of a glacier avalanching into a narrow fjord in Lituya Bay, Alaska. The 1958 avalanche had been triggered by a magnitude 8.0 quake 13 miles away on the Fairweather fault. With hindsight, scientists now suspect that landslides play a much greater role in tsunami generation than was earlier believed.

Of all the 1992-1998 events, the Papua New Guinea event was by far the most conspicuous and anomalous. The unprecedented scale of the PNG tragedy, coupled with the unexpectedly large size of the tsunami and the complexity of the local geology, inspired an intense international scientific effort to assess what had triggered the event. Recent evidence now suggests that the 1998 Papua New Guinea tsunami, as well as almost one-third of those in the 1990s, were generated by quake-induced landslides. Consequently, federal agencies such as the National Oceanic and Atmospheric Administration (NOAA) and the U.S. Geological Survey have been quietly assessing the landslide potential of near-shore submarine shelves and canyons.

There are several important differences in the character of tsunamis triggered by landslides and those triggered by earthquakes. Tsunamis generated by earthquakes tend to have longer wavelengths, longer periods, and a larger source area than those generated by mass movements of earth. And, in most cases, the size of a tsunami triggered directly by an earthquake and the deformation of the sea floor is limited by the magnitude of the event. On the other hand, a tsunami generated by a landslide can grow much larger; its size is limited only by the amount of vertical motion in the triggering flow of earthen debris.

When a potential tsunami-triggering earthquake occurs, sufficient information is often available to predict whether or not a massive wave will be created. However, landslides often occur unexpectedly, and they can strike anytime after even a mild earthquake. There are at least four characteristics of a landslide that determine whether or not a tsunami will form, including its length, width, thickness, and the inclination of the slope that fails and triggers the mass movement. Unfortunately, none of these characteristics can yet be accurately predicted, and the relevant information typically comes only after the event.

The potential for tsunami-triggering mass movements can often only be assessed in the context of the surrounding geology, and this typically requires a multidisciplinary approach to modeling and costly ocean cruises. During January and February of 1999, lengthy cruises were undertaken to map the bathymetry off the north coast of Papua New Guinea to determine what had caused the 1998 event. In other instances, information on submarine landslides can be obtained due to resultant breaks in underwater cables. The timing of cable breaks and the subsequent loss of communications can be used to document the extent of the causative submarine slump or flow and actually record its timing. Fishermen at sea during an undersea debris flow have also reported seeing muddy seawater rolling toward the surface. In the absence of such information, it is nearly impossible to determine the exact timing, let alone the nature of submarine landslides.

Tsunami Modeling

State-of-the-art computer simulations of tsunamis are now helping scientists gain a more comprehensive understanding of how tsunamis form, travel, and strike the shore, thus allowing more accurate and timely warnings as well as better risk assessments (*color plate 16*). Essentially, tsunami models can be thought of as the synthesis of earthquake, wave, and flood models, all of which require certain initial inputs. To model a quake-generated tsunami, the first information required is the size and distribution of the sea-floor deformation and the amount of strain energy released. The amount of sea-floor deformation can either be measured or predicted by a different type of computer model. Obviously, the most accurate means of determining sea-floor deformation is to actually measure it; this is done by comparing the underwater topography (bathymetry) of the sea floor, post-quake, to previously collected pre-quake data. Unfortunately, this is not a fast (nor sometimes even possible) prospect.

A quake's deformation area is generally defined as a rectangular box enclosing the epicenters of its aftershocks. So to measure the deformation area, seismologists may have to wait for days after the main quake for the aftershocks to stop. A minimum of one week following an earthquake is generally needed before deformation mapping can be accomplished. Even if it were possible to somehow immediately determine the entire deformation area, a bathymetric survey would be needed to measure the extent of deformation within the region, and this could take days to weeks to conduct. New equipment using water-penetrating radar from a helicopter, currently under development by the U.S. Army Corps of Engineers, will hopefully allow for more rapid mapping of the sea floor. But this new technology has yet to be used after a tsunami event, and it still would take probably a couple of days to complete a sea-floor survey.

More importantly, in most regions of the globe, there is only sparse or rough pre-tsunami data to compare to. In fact, we know more about the surface of the Moon and even Venus than we do about the sea floor, which happens to comprise nearly three-fourths of the surface of our own planet. Undoubtedly, this lack of information is a reflection of national priorities in research spending. In the bathymetric survey after the deadly 1998 Papua New Guinea tsunami, the sea floor was mapped with a horizontal resolution of a less than a meter. Unfortunately, the accuracy of available bathymetric data pre-tsunami is on the order of tens to hundreds of meters, making comparisons of before and after conditions rather crude. Without detailed knowledge of the pre-tsunami sea floor, it is much more difficult for scientists to determine whether change has occurred, and if so, when it happened—before, during, or after a tsunami event.

Given the difficulties in actually measuring sea-floor deformation, scientists usually rely on seismologic model predictions. In other words, some of the input to a tsunami model often comes from an earthquake model. These computer simulations are based on models of elastic deformation and are applicable for all earthquakes, whether submarine or not. For each earthquake, seismologists estimate what the surface deformation would be in a material with the same elastic properties as the Earth's interior, given an internal displacement of the size inferred from seismic records. This approach provides an adequate estimate for the ground deformation; and when the earthquake epicenter is inland, the predicted displacement can be checked against measurements from topographic surveys.

Each earthquake event is modeled using what is now referred to as the Harvard fault plane solution. The Harvard solution uses seismic data from different stations around the world and then determines what type of fault could have generated the seismic waves at its hypocenter. The solution reveals, fairly accurately, the energy released by the earthquake. Seismologists then estimate a preliminary size of the deformation area and the vertical ground displacement. Other earthquake characteristics important for estimating the pattern of ground deformation are inferred from the known structure of the fault area. For example, a subduction zone usually dips at a known angle; one plate is subducted under the other at a constant angle. If the particular fault has ruptured before, then the orientation of the fault with respect to the coastline is usually known. One parameter that is hard to determine is the actual slip angle, but this can often be estimated from the seismograph recordings.

Tsunami models use the energy released, the size of the deformed area, the mean vertical displacement, and the dip, strike, and slip angles to infer a large-scale pattern of sea-floor deformation. Then, the models assume that water displacement occurs instantaneously and that the initial tsunami wave has the same shape as the sea-floor deformation. In other words, the water develops the same hills and valleys as the deformed sea floor beneath it. This is a reasonable assumption, based on the principle of mass conservation. Whatever mass of fluid is displaced by the sea floor moving up or down causes an equivalent displacement of the water in the same direction. The difference is that while the sea floor stops moving at the end of the earthquake, gravity causes the water surface to attempt to restore itself, thus triggering the wave motion. Now, the initial wave conditions have been established, and the model must change gears to simulate how the tsunami travels through the open ocean.

The evolution of the tsunami from its source to the target coastline is simulated using wave-modeling techniques. The initial wave conditions, the depth, and the underlying sea-floor topography are used as inputs. When the simulated wave arrives at the coastline, the wave model is linked to an inundation or flood model, which determines the properties of the tsunami as it moves onshore—a real computational challenge.

Up until the 1990s, reliable computer models for estimating coastal inundation did not exist. As in many other scientific disciplines, in the absence of good models, tsunami understanding advanced in the United States primarily through laboratory experiments and theoretical studies—some that have now proven incorrect.

Laboratory Simulations

Of course scientists cannot create real tsunamis in the laboratory, so they produce waves that have proportions similar to a real tsunami. Small-scale waves are generated in an enclosed water basin or tank having the correct proportions in terms of their wavelength, height, and depth. The process is similar to that used by aerospace engineers to test small-scale models in a wind tunnel. Using a scale model of 1:100 means that all the dimensions of the model must be 100 times smaller than the real thing (tsunami, plane, rocket, etc). To simulate a real tsunami generated in a depth of 100 meters with a height of 1 meter, the wave in a laboratory basin with a water depth of 0.5

meter would need to be 0.5 centimeters high—it would hardly be visible. Scientists have found that in the laboratory, water waves smaller than about 6 centimeters are not good replications of a tsunami because they are disproportionately affected by surface tension, which acts on small-scale fluid motions like soap bubbles. But real tsunamis are only affected by gravity and not surface tension, so a wave larger than 0.5 centimeters must be used to obtain meaningful results. Now imagine trying to model a 1-meter-high tsunami generated at an ocean depth of 2,000 meters. Using a 1:100 scale model, a huge tank would be needed, some 20 meters (65 feet) deep. Worse yet, there is another tsunami characteristic that needs to be modeled correctly—the wavelength. A typical tsunami in the open ocean might have a wavelength of 100 kilometers; using the 1:100 scaling, the model wave would have to be 1 kilometer long!

In addition to needing a gigantic laboratory facility, it would take an enormous amount of energy to generate a 1-kilometer-long wave in 20 meters of water. Furthermore, if the distance from generation of the tsunami to the coastline is on the order of 300 kilometers, then the laboratory tank would have to be at least 3 kilometers long to model correctly the propagation of the tsunami from its source to target. One could try using a different scaling ratio, maybe 1:1000. This would make the dimensions of the tank manageable, for it would require a 2-meter-deep tank that is at least 3 meters long. However, the wave would then be only 1 millimeter high, hardly noticeable, hardly measurable, and surface tension would make interpretation of the results nearly impossible.

Thus, to address this problem, scientists must generate tsunamis in the laboratory that are proportionally much larger than are typically found in nature. For example, in a tank 50 centimeters deep, researchers generate a 10-centimeter wave. The equivalent in real tsunami terms is a 20-meter (65-foot) wave at an offshore depth of 100 meters, a wave larger than even filmmakers create. Although this approach has merit for model validation—any model that would work for a 20-meter tsunami would work for a 2-meter tsunami—for years it led the tsunami community down the wrong path. Because few field measurements were available, scientists assumed that tsunami waves in nature were as big as their laboratory versions implied, and this led to incorrect conclusions about how tsunamis behave.

The tsunami events since 1992 have helped identify where the mistakes were made and have suggested how to improve the earlier flooding models. It was fortuitous that so many tsunamis happened in relatively recent years, just when inundation models were beginning to produce repeatable results. In both Japan and Russia, scientists were developing and beginning to test what were then crude tsunami inundation models. In the United States, Dr. Eddie Bernard, currently the director of NOAA's Pacific Marine Environmental Laboratory (PMEL) in Seattle, Washington, began pressing the U.S. scientific community to develop reliable tsunami flooding models for better evacuation planning. At the same time, funding through the U.S. National Science Foundation started to provide the required resources.

Models using older technology had previously been used to produce evacuation maps, but they could not calculate inundation. As with the Japanese models, the older Soviet and American models could only predict how tsunamis travel from deep to shallow water, but they could not simulate movement onto the shore. Since estimates of inundation are critical to evacuation planning, the pre-'90s practice was to make predictions based only on wave height offshore. In essence, the early models treated the ocean as if it had vertical walls near the coastline, and predicted the height of the tsunamis onshore as they hit these seawalls. Scientific research later proved that this was an unsatisfactory method, since tsunami height can change dramatically when it comes ashore. Recent work at the University of Southern California has now demonstrated that the profile of the coastline is a critically important factor in tsunami inundation.

In 1990, with funding from the National Science Foundation, scientists from the University of Southern California, Cornell, and the University of Washington developed a comprehensive research program to conduct large-scale experiments to help provide lots of laboratory data for model validation. The real difference with respect to earlier laboratory investigations was the size of the experimental models. Scientists conducted a series of experiments in a large wave basin at the U.S. Army Corps of Engineers Coastal Engineering Research Center (CERC) in Vicksburg, Mississippi. In the CERC water tank, which is 30 meters wide, 30 meters long, and 60 centimeters deep, realistic waves could be created by a horizontal wave generator with sixty different paddles, each moving independently. It was here that the 1992 Babi Island tsunami described earlier was modeled. The model revealed that the wave had split in two, traveled around the island, and produced massive run-up on the island's normally protected backside.

Models that combine our understanding and ability to simulate earthquakes, waves, and coastal inundation are continually improving. When coupled with better mapping of the sea floor, particularly in the coastal zone, and advances in computer technology and improved validation experiments, modeling of tsunamis promises to improve our ability to assess and predict the deadly impacts of these monster waves.

Local vs. Distant Tsunamis

Scientists often divide tsunamis into two types, local and distant. Local tsunamis are typically referred to as nearfield, and distant ones as farfield, remote, or tele-tsunamis. This is a relative means of distinguishing tsunamis based on the distance from their origin to where they strike the coast. One person's local tsunami can be another person's distant tsunami. The tsunami triggered by the 1964 Good Friday quake in Alaska was a nearfield tsunami in Alaska, but a farfield tsunami in Crescent City, California. The 1993 Kuril Island tsunami was a local tsunami in Shitokan Island, Japan, but a distant tsunami in Hawaii. In general, a nearfield or local tsunami in Japan is a farfield or distant tsunami in Hawaii, and vice versa.

Until the devastating tsunamis of the 1990s, scientists thought that ground shaking was always a precursor of local tsunamis. If an earthquake occurred close to the shore and was large enough to trigger a tsunami, people living near the coast should feel the ground shaking (and could be warned to move to higher ground). On the other hand, distant tsunamis may originate so far away that the causative earthquake may be imperceptible to all but the most sensitive seismic instruments.

Scientists also thought that since there were few reports or data from local events, they could understand all types of tsunamis from the study of documented distant tsunamis. If you can model a farfield tsunami, you should be able to model a nearfield tsunami—after all, they are the same phenomenon, just viewed from different parts of the ocean. This way of thinking changed dramatically on September 1, 1992, when a 6-meter (20-foot) tsunami hit the Pacific coast of Nicaragua without any precursory ground shaking. The tsunami advanced almost a mile inland and killed 200 people. Yet the Nicaraguan tsunami was a local tsunami (generated by a quake relatively close by), and, according to theory, it should have been preceded by ground shaking. The scientific community was perplexed and motivated to better understand tsunamis. After all, the thought of a towering tsunami striking the shore minutes after it is generated, with no warning signs, is a nightmare. Imagine if a local tsunami struck Southern California during a weekend afternoon in the summer, with thousands of people at the beach and little or no warning. This concern was reinforced by the 1998 tsunami in Papua New Guinea, in which the ground shaking was so mild that many residents were busy assessing the damage to their homes, unaware of the wall of water that would soon sweep away their houses along with those still inside. Clearly, not all tsunamis, not even locally derived tsunamis, warn of their dangerous approach.

The 1992 Nicaraguan and the 1994 East Javan events proved puzzling in another way: the magnitude of the tsunamis was far greater than expected based on traditional seismological models of sea-floor deformation. As it turned out, both tsunamis were caused by what scientists now refer to as a "silent" tsunami earthquake. These tsunami earthquakes are both silent and slow. They do not cause extensive ground shaking in nearby areas, and the associated fault rupture occurs very slowly. Yet they are surprisingly efficient at generating tsunamis.

Professor Hiroo Kanamori of Caltech first proposed an explanation for why these earthquakes are so effective at producing tsunamis, but neither he nor anybody else expected these events to be so prevalent. If a fault ruptures slowly, at a speed coinciding with the speed of waves in the overlying water, then an earthquake appears to be particularly effective at triggering a tsunami. Dr. Emile Okal of Northwestern University and his colleagues recently developed a new method to determine, in real time, whether any given earthquake is likely to generate a once-unexpected tsunami. The method is called TREMORS (Tsunami Risk and Evaluation Through Seismic Moment for Real time System) and is now used in the tsunami warning center that protects French Polynesia in Pappeete, Tahiti.

Even after the slowness of the triggering earthquakes had been accounted for, the tsunamis that struck the shorelines of Nicaragua in 1992 and East Java in 1994 were mysteriously devastating (*color plate 21*). Scientists began to consider the possibility that the near-shore bathymetry may have influenced how the tsunamis had hit the shore. At the time, there was growing speculation that near to shore, a region's undersea topography could play a more important role than previously believed. While this seems logical, especially when we consider how a submerged reef or sand bar can cause waves to break and create excellent surfing spots, it goes against conventional wisdom among tsunami scientists. They had always argued that tsunami waves are so long that the details of the near-shore sea floor are unimportant; to a tsunami the shoreline acts like a vertical wall (remember the early models that treated the coastlines as walls).

One of the puzzling aspects of the damage in the 1992 Nicaraguan tsunami was that along the shore the intensity of destruction was highly irregular. In some regions little damage was done—even the beach umbrellas were left standing— whereas other sections of the shoreline were completely decimated. In 1995, scientists from USC, the U.S. Army Coastal Engineering Research Center, and the University of Washington went back to Nicaragua to carefully measure the offshore bathymetry. Using an old Nicaraguan navy vessel, they mapped the sea floor starting from one mile out all the way to the shoreline. It was a difficult task, as the boat had to traverse uncharted waters and at times go blindly into dangerously shallow water less, than 1 meter deep. The survey revealed that at about 200 to 600 meters offshore, most of the Nicaraguan coast is fronted by a submarine coral reef, and that openings in the reef coincided with the greatest amount of tsunami damage along the shore. Where the reef was continuous and intact, damage had been minimal. Openings in the reef had allowed the tsunami to pass through, whereas regions of solid reef dissipated or altered the course of the huge waves. Tsunami experts now agree that the underlying structure of the sea floor fronting the coastline, often within what is called the surf zone, can greatly influence how tsunamis strike the shore. The destructiveness of the tsunami that wreaked havoc on the Nicaraguan coast in 1992 can now be explained by a slow, silent tsunami earthquake and the magnifying effects of the nearshore bathymetry.

Tsunami Prediction

If reliably predicting earthquakes is difficult, then reliably predicting tsunamis is nearly impossible. The prediction of both earthquakes and volcanic eruptions is generally based on probabilities, which in turn are based mainly on how often such events have happened in the past. For tsunamis, this sort of probability prediction is even more difficult. Not only do we need to know the probability of a given fault rupturing, but also whether it will produce a tsunami during a particular event. And tsunamis are much rarer phenomena than earthquakes. On a global scale, a magnitude 6 earthquake or larger occurs at least once every week. A comparatively large tsunami was, for many years, thought to occur only once every decade. Since 1992 there have been ten large tsunamis, but even so, based on the 1992-1998 period, a large tsunami could be predicted to occur only once every five months or so. Not only do tsunamis occur relatively infrequently, but their historic record is both short and incomplete.

Even in Japan, where tsunami documentation extends back 1,000 years, it is extremely difficult to predict the likelihood of a tsunami based on the available records. Tsunami descriptions contain varying amounts of detail, and it is only in the last 100 years that accurate wave records exist and have been correlated to specific earthquake events. Also, the landmarks referenced in reports have often changed or are now nonexistent. It is difficult to interpret a historic report that documents a tsunami's flooding relative to the house of a prominent citizen when that house (and citizen) no longer exists. The local conditions along the coast can also change. In effect, many of the past estimates of coastal inundation are meaningless, except for providing a record that an event did occur and a qualitative estimate of its size. Nonetheless, records and local lore enable residents to infer how often a tsunami is likely to occur. As in assessing earthquake risk and building quake-resistant structures, tsunami records are useful mainly to convince people of the hazards and to find ways of reducing the risks.

What scientists do know is that a certain class of faults will invariably generate tsunamis if the rupture area during an earthquake is large enough. Where earthquakes have previously produced tsunamis, particularly on thrust faults within a subduction zone, they will probably do so again. Before the 1994 Mindoro, Philippines, earthquake, strike-slip faults were not supposed to trigger tsunamis, yet here a fairly moderate strike-slip event generated a sizable tsunami. And before 1998, earthquakes with relatively small to moderate moment magnitudes were considered unlikely to generate tsunamis, yet Papua New Guinea was hammered by a 15-meter wave after a small-to-moderate earthquake. Far less is known about the probabilities of tsunami-triggering mass movements, such as submarine landslides, being generated by an earthquake. Even in the densely populated coastal regions of the United States, landslide hazards have only recently been recognized. A map of the sea floor off Santa Monica Bay recently released by the U.S. Geological Survey reveals evidence that in the past large mass movements have occurred in the bay, suggesting that they may well happen again in the future.

Nonetheless, if we know the probability of a given fault rupturing, it is possible to estimate the likelihood for a tsunami and where and how it will strike. However, at any one locale, we must estimate the probability of all possible events, considering all faults within the area, and all segments of all faults. This process is incredibly time-consuming, for each computer simulation takes hours, and thousands of runs are needed. In the end, the final inundation or strike probability is only as accurate as the earthquake probability. Consequently, evacuation planning tends to rely more on the highest possible wave that may strike a given area, versus the probabilities of all potential events. Because the record of tsunami events is relatively short, it remains difficult to predict the likelihood of a tsunami in any one region.

Paleotsunamis and Asteroid Impacts

Over the last twenty years, a new method has evolved to investigate tsunamis and to estimate how often they occur—the study of paleotsunamis. The term *paleotsunami* refers to ancient tsunamis, those that occurred before the written record. When tsunamis come ashore, they often carry sediments from the sea floor. After the event is over, water returns to the sea, but much of the sand and mud are left behind. Tsunami deposits often contain the shells of small marine creatures, such as diatoms or foraminifera, whose origin can be easily traced to the sea, or they may be distinctly chaotic piles of debris and rocks. Over time, tsunami sediments typically become buried under layers of debris weathering off the land. By digging a trench or looking for exposed sedimentary beds, a geologist can uncover tsunami deposits and study ancient events. Because most paleotsunami deposits contain wood and other organic debris, they can be dated fairly accurately using radiocarbon dating techniques. In fact, it is more difficult to date recent deposits than ancient ones because most standard dating methods can place an event only within a range of about 200 years.

The discovery of a paleotsunami deposit in Texas led one group of scientists to a very interesting and exciting conclusion. A group of sedimentologists, led by Dr. Jody Bourgeois from the University of Washington, found marine sediments far inland. Yet at the time the sediments were deposited, there was no evidence that the sea had ever extended that far inland in Texas. The researchers concluded that the only explanation for the deposit was that it was a paleotsunami deposit. Further evidence suggested that only an asteroid impact in the Gulf of Mexico could have possibly generated a big enough tsunami to do the job. Dating of the paleotsunami deposit indicated that the impact occurred about 65 million years ago—the same time as the demise of the dinosaurs.

It now appears that the asteroid which created the Texas tsunami deposit may have been the same one that led to the extinction of the world's dinosaurs. In fact, a few years after the discovery of the Texas deposit, oceanographers discovered a huge submarine impact crater off the Yucatan Peninsula in the Gulf of Mexico. The movie *Deep Impact*, which dramatically recreated an asteroid-triggered tsunami striking the eastern United States, was based on this wonderful investigative work. In a similar case, a paleotsunami deposit led Dr. Wylie Poag of the U.S. Geological Survey to another impact crater, this one 50 million years old, beneath the mouth of Chesapeake Bay. While asteroid-triggered tsunamis do occur, the infrequency of such events is such that few of us take the risks seriously; and even if we did, what could be done to mitigate such titanic hazards?

Tsunamis and the Pacific Northwest

Over the last two decades, U.S. scientists have become increasingly concerned about the potential threat of a major earthquake and tsunami within the Cascadia Subduction Zone, just off the coast of Oregon, Washington, and northernmost California. Here, the North American and Pacific Plates meet, and a smaller plate, the Juan de Fuca Plate, is being squeezed under. Because records prior to the 1770s are absent, little was known about the region's seismic and tsunami history. Among scientists, controversy reigned about the mechanics and size of potential earthquakes in the region. Some geologists argued that the subduction zone ruptures slowly and produces only moderately sized earthquakes, tremors that are unlikely to produce tsunamis. Other scientists argued that once every several hundred years the fault within the subduction zone could rupture suddenly and completely. A less frequent but great quake was a much more alarming scenario, because hundreds of thousands of people live along the shores of the Pacific Northwest. If a large enough tsunami were triggered, it could enter Puget Sound and wreak havoc on Seattle, travel down the coast to California, and then cross the Pacific to Hawaii and Japan. On the other hand, if the other theory was correct and the fault zone ruptures more often, the resulting earthquakes would probably be smaller and pose less of a threat. But without historic records, there was no way to determine which scenario was more likely to occur.

In the 1980s, a group of scientists from a variety of fields began to look for clues that would shed light on the history of earthquakes and tsunamis in the Pacific Northwest. Geologists from the U.S. Geological Survey and several universities began to search for geologic evidence of large earthquakes. During their investigations, U.S. Geological survey geologist Brian Atwater discovered sedimentary layers on Washington's outer coast that are similar to a sequence of deposits that resulted from great earthquakes in Alaska and Chile. His findings suggested that the Cascadia Subduction Zone had produced large earthquakes in the past, and that some had generated tsunamis. Soon thereafter, geologists began to find further evidence that a great earthquake and giant tsunami had hit the coast of Washington and Oregon some 300 years ago, between about 1680 and 1720.

By the early 1990s, there was no question of the seismic potential of Cascadia, only a question of when and how big. Would a rupture in the subduction zone trigger a giant magnitude 9 quake, or a series of smaller but still significant events, say magnitude 8? Then learning of the evidence for a great quake in the Pacific Northwest some 300 years ago, geophysicists Kenji Satake, Yoshinobu Tsuji and Kazue Ueda began to search for evidence of a corresponding distant tsunami in Japan. Records in Japan revealed an unusual tsunami on January 27, 1700 and it was an orphan; scientists were unable to associate it with any local earthquake, and the inundation area reported in Japan was more characteristic of a distant than a local event. Having noted that tsunamis of similar magnitude had hit Japan coming from across the Pacific, they looked for similar evidence of a large tsunami occurring at about the same time in the Kamchatka Peninsula, and in North and South America. In coastal Washington, they discovered an Indian legend that told of a large wave hitting the coast on a winter night in the not so distant past. Was this also a consequence of the 300 year old giant earthquake in the Cascadia Subduction Zone ? A tsunami triggered on the night of January 26 could have struck Japan's shores the following morning. Scientists then asked how large an earthquake in Cascadia could produce a large tsunami in Japan. The answer: a mammoth quake, possibly magnitude 9.0, in the Cascadia Subduction Zone. The sediments, the Japanese records, the modeling data, and even the Indian legend all tell the same story. A major quake tore through the Cascadia Subduction Zone and triggered a massive tsunami around January 1700.

Evidence now indicates that the Cascadia Subduction Zone can indeed produce massive earthquakes and trigger tsunamis, possibly once every 300 to 400 years. Therefore, in all likelihood, when the next great quake strikes, it will probably generate a substantial tsunami. Dr. Eric Geist at the U.S. Geological Survey and others are now working to predict just how big that tsunami will be. Others in the school of gloom and doom have suggested that a fault rupture in the Cascadia Subduction Zone could also awaken one the sleeping volcanoes in the region, such as Mount Rainier, and trigger an eruption. This suggests a catastrophic scenario seemingly too outrageous even for Hollywood scriptwriters, a giant quake followed minutes later by an enormous tsunami, followed hours or days later by an explosive volcanic eruption.

If a great quake were to occur within the Cascadia Subduction Zone, coastal regions along the U.S. West Coast could be slammed by a series of monster waves within minutes. The Federal Emergency Management Agency (FEMA) estimates that the double whammy of an earthquake and tsunami in the Pacific Northwest could cost $25 to $125 billion in losses. The tsunami alone could account for losses totaling $1 to 6 billion. And between local residents and tourists, over half a million people could be threatened by such an event.

Tsunami Impacts and Reducing the Risks

Tsunamis present a clear and deadly threat to coastal populations, particularly around the rim of the Pacific Ocean (*Figure 4.3*). Japan and Russia both have a long history of tsunami impacts. In the United States, Alaska, California, and Hawaii have been well recognized as areas at risk for tsunamis. Recent evidence suggests tsunami hazards also exist in the Caribbean and along the shores of the Gulf of Mexico, and are of particular concern in the Pacific Northwest. We cannot stop tsunamis from occurring any more than we can prevent volcanic eruptions and earthquakes, but we can try to reduce the risk to those living in harm's way.

To reduce risks, several aspects of tsunamis must be considered: the shortness or absence of a warning, a potentially long duration, and the extent and force of flooding in any given region. As discussed already, ground shaking, a loud bang, or a rapid drop in sea level may precede a tsunami, or it may strike without precursory signs. Tsunamis may also pose a threat for a dangerously long time during events.

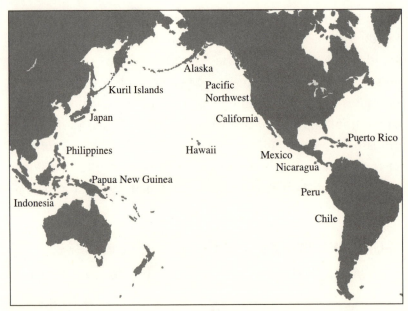

Figure 4.3
Regions where tsunamis have struck in the past and that remain at risk.

Earthquakes can be devastating, but once the mainshock is over, the subsequent aftershocks tend to be weaker, so search and rescue efforts start immediately. Tsunamis generated within 80 to 160 kilometers from shore can strike in less than an hour, but additional waves can arrive much later. And when a tsunami enters a bay, it will likely set up water motions or oscillations that can last for days. The tsunami generated from the eruption of the Krakatau volcano in 1883 set the global ocean in motion for several days; related waves were recorded at Greenwich Observatory in England for a week after the actual event. Following a tsunami impact, it is often difficult for search-and-rescue personnel to plan their missions, for it is impossible to know whether or when a second or third wave is going to crash down. It is for this reason that advance planning and education is so important for any place that faces a risk of tsunamis.

Tsunami impacts in poorly developed regions are often worse in terms of human tragedy and receive less attention than similar disasters in well-developed countries. Higher losses are often due to a lack of understanding and preparation. Less attention is usually afforded tsunami disasters in these regions simply because we may not even know that they have occurred, and they tend to result in relatively lower economic costs. During the 1992 Flores Island tsunamis, the nearby island of Babi was completely destroyed, and out of a population of 1,000, 800 died. The waves advanced inland sweeping every single structure off the beaches of Babi and depositing them in a coconut grove at the foot of a hill, about a quarter mile from the coastline. Eyewitness reports told of extremely gruesome scenes, with human remains impaled high in the trees and dangling for days. Even weeks after the event, when the International Tsunami Survey Team conducted its post-event investigation, there were human remains buried in the rubble. Right after the tsunami hit, the disaster relief officials in Maumere, the capital of Flores Island, were not even aware of the Babi catastrophe. The tsunami had swept all the boats away from Babi, and survivors had to swim 8 kilometers across to Flores to report what had happened. It took two days for disaster relief efforts to reach Babi. After the disaster, the government prohibited rebuilding on the island, but new fishing settlements have since cropped up in the very area that was destroyed.

Compare this to a tsunami that struck Japan on July 14, 1994. The tsunami and resulting fires killed more than 200 people. The Japanese government then spent $600 million to rebuild Okushiri Island, a sum that amounts to $600,000 per person for a community of 1,000 people. At least $400 million could have been saved by relocation of the families in Okushiri to the mainland of Hokkaido. But Japanese society was so shocked by the magnitude of the disaster that for a while it seemed money was no object when it came to restoring Okushiri to its pre-tsunami condition. Even the recovery from the 1995 Kobe earthquake that affected millions of people did not cost as much money per person as that from the Okushiri tsunami. And although a tsunami-prone region may be rebuilt, the risk of future disaster remains. Was the Babi or Japan tragedy more expensive? How can we put a price on human suffering and compare economic costs in such vastly different cultures? Unquestionably, no matter where and how many people are involved, we must try to reduce tsunami risks.

The easiest and most effective tsunami countermeasure is to teach people living, working, and recreating along the shore to expect and recognize the warning signs of a tsunami. When the ground begins to shake along the shore, the message is clear—move inland and up, as fast as you can. It is also important not to panic, because there should be time to safely move to high ground (where possible). No tsunami, not even a locally generated tsunami, has struck in less than five to ten minutes after a triggering event.

Anyone enjoying a seaside day at the beach should also know that unusual water motion, such as the rapid retreat of the water, or an unusually loud bang (without a source) are tell-tale signs of a mounting wave and an indication to start moving inland. When evacuating, it is best to quickly climb to high ground and to remember that roads and bridges are often destroyed by associated ground shaking. Tsunamis also advance further and faster along flat surfaces. And if caught in a tsunami, watch for flying debris, and grab anything that floats.

It is equally important to realize that tsunamis are natural phenomena of long duration. One should not only move away from the shore but stay away until one hour later, or preferably when civil defense authorities, if close by, give the all-clear sign. Tsunamis are notorious for being unpredictable; there is no way at present to know whether the first, second, or third wave will be the most devastating. Hilo, Hawaii, in 1946 was struck by a total of nine colossal waves, originating from the April Fool's Day earthquake in the Aleutian Islands. Many perished because they did not evacuate far enough after the first wave. Some survivors were picked up floating in the open ocean some twelve hours after the first wave hit.

In the poorly developed and remote coastal regions of the world, where tsunamis are relatively common, the International Tsunami Survey Team recommends that each family have a rapid vertical evacuation plan. In other words, choose a tree and carve steps into it, or have a ladder handy for fast climbing. Trees with a deep root system (well anchored) and a slender trunk (low resistance to water flow) often survive tsunami impacts. Upon recognizing the signs of a tsunami or seeing a monster wave offshore, family members can climb up the tree as far as possible. The team also recommends that in areas of the world where coastal villages are far from warning centers, each community should have a designated tree with a warning bell about 9 meters (30 feet) above the ground. If anyone in the village sights a tsunami offshore or feels an earthquake, they can climb up and ring the bell to warn everyone to move away from the water. In more developed regions, if time is limited, the best choice is to move to the top of a two- or three-story building. The construction of monuments to victims of earlier tsunami disasters is also recommended, so that residents will recognize and be reminded of the serious risks involved. This may seem trivial, but people tend to have short memories. When the International Tsunami Survey Team visited Nicaragua three years after the catastrophic tsunami in 1992, none of the survivors they had met previously were anywhere to be found; all had been relocated by the government. People now living within the villages and towns rebuilt along the shore had no memory or appreciation of the dangers posed by tsunamis.

In some regions of the world, the government has sought to reduce the risk of tsunamis by building protective structures. The Japanese have built seawalls 9 meters (30 feet) high to protect communities at risk. A large proportion of the $600 million spent following the 1993 Okushiri event went toward the building of an extensive seawall. Along the southern coast of Honshu, the central of Japan's three big islands and the one on which Tokyo is located, numerous seawalls have been constructed to protect coastal towns and fishing villages. The structures resemble the huge walls of a medieval castle, with imposing steel gates. In the event of an impending tsunami, a dedicated crew of volunteer firefighters slowly close the gates. Each gate is equipped with a small trap door to allow for late-coming evacuees.

In the more tropical parts of the Pacific, seawalls are out of the question. Imagine a small fishing village in Indonesia or the Philippines surrounded by a high concrete or metal wall. Even if it were built by the government, fishermen would find a way to move their homes in front of the seawall to get closer to the water. This has happened in Japan, and the government responded by building a second smaller seawall in front of the first. Or imagine the popular and profitable tourist beach of Waikiki in Honolulu, Hawaii, fronted with a massive 9-meter seawall—a highly unlikely scenario. In the United States and elsewhere, the preferred method for tsunami protection is zoning. Hilo, Hawaii, has been hit by tsunamis twice this century; consequently the city has banned construction in the area that was inundated during the previous events. The area is now a park that houses a tsunami memorial. All the hotels and other buildings in tsunami-prone areas are designed for vertical evacuation and to withstand flooding. These structures lack solid walls at their base, so that a tsunami can pass beneath and through the building with minimal resistance and minimal damage.

Inundation mapping can also help reduce a region's risk from tsunamis through better response and city planning. Many visitors to Hawaii notice the inundation maps printed in the telephone book under "useful information." These maps show the coastal areas at highest risk. Residents in these areas should evacuate immediately if a warning is issued, regardless of whether they notice precursors. Similar maps are being developed for all U.S. states bordering the Pacific Ocean. The effort has been spearheaded by Drs. Eddie Bernard and Frank Gonzalez of NOAA, who have lobbied Congress and set up the National Tsunami Hazards Mitigation Steering Committee. Inundation maps are prepared mostly based on modern tsunami-modeling techniques developed by NOAA and USC. These maps will be used to educate local communities about tsunami risks and help emergency personnel plan for operations in the event of a tsunami. The Los Angeles County Fire Department is becoming a world leader in tsunami response planning and is studying search-and-rescue methods that may be needed specifically in the aftermath of a tsunami. Inundation maps are also helping city planners prepare for some of the other dangers posed by tsunamis. For example, Southern California has numerous world-renowned beaches, and many, such as those in Santa Monica, Venice, Long Beach, and Orange County, have parking lots to facilitate beach access. Unfortunately, cars in a beach lot are a serious tsunami hazard; even small tsunamis can turn a car into a dangerously fast, heavy, and deadly missile. Inundation maps also help officials decide where to relocate schools and critical facilities to safer areas and plan for safe evacuation routes. For many areas, an early warning system may provide one of the most effective means of reducing risk, particularly for distant tsunamis.

Warning Systems

Tsunami warning systems are designed to alert coastal populations that a potentially disastrous series of monster waves is approaching. Ultimately, a tsunami warning system would operate and provide information in a way similar to the now-familiar and very effective system for hurricane warnings.

There are four major tsunami warning centers in the Pacific: the Alaskan Tsunami Warning Center near Anchorage, the Pacific Tsunami Warning Center near Honolulu, the French Polynesia Warning Center in Papeete, and the Japanese network in Japan. The Russians also operate a warning center on the eastern coast of Siberia, although this particular system is not always operational. At present, early warning centers are only capable of issuing a timely alert for distant tsunamis. A more modest system specifically designed for local tsunamis operates in Chile with technology provided by NOAA. By and large, however, locally derived tsunamis leave little time for detection or warning, partially because of our inability to detect tsunamis in the open ocean; however, this situation is starting to change.

Warning centers have traditionally issued alerts based on potential tsunami-generating earthquakes and historical data. Seismic stations throughout the world constantly transmit earthquake data to the regional tsunami warning centers. At the centers, the likelihood of tsunami formation is then assessed based on the location and magnitude of earthquakes, and in some instances the past history of tsunami generation in the area. If a tsunami is deemed likely, a warning is issued for coastal areas that could be effected within several hours or less, and a watch is issued for regions that may be impacted later. At the Alaska Tsunami Warning Center, regional warnings are issued within fifteen minutes of an earthquake and are based solely on seismic data. Any coastal earthquake over magnitude 7.0 results in an immediate tsunami warning. Warnings outside the Alaska Center's region are coordinated with the Pacific Tsunami Warning Center in Hawaii. There, a warning is issued based on seismic data combined with historical records and data from coastal tide-gauge stations.

With this warning system, there is no way to confirm tsunami predictions, and this has resulted in numerous false alarms. Unfortunately, false alarms can cost millions of dollars and, as in the story of the boy who cried wolf, erode the public's trust in those issuing the warnings. Since 1948, twenty tsunami warnings and evacuations have been carried out in Hawaii. Of the twenty warnings, fifteen were considered false alarms. For instance, during the magnitude 8.0 Kuril Island earthquake in 1993, a 10-meter tsunami slammed into and devastated the southern coast of Shitokan Island, north of Japan. Because the area lies adjacent to a subduction zone deep underwater, it is notorious for generating large tsunamis. So the Pacific Tsunami Warning Center issued an evacuation warning for the Hawaiian Islands three hours before the waves were estimated to arrive. Fortunately, when the waves arrived, they were less than 0.3 meter (1 foot) high. However, conservative estimates indicate that the false alarm and evacuation cost the Hawaiian economy roughly $30 million. But the costs of not evacuating or returning to an evacuated area too soon are clearly much greater. The tsunami triggered by the 1964 Good Friday earthquake in Alaska killed 159 people in Hawaii. In contrast, in the town of Crescent City, California, an evacuation order was issued before the first wave arrived, and many lives were saved. Unfortunately, some people returned to the evacuated area too soon, and eleven people perished due to late-arriving waves.

To combat the problem of false alarms, confirm tsunami warnings, and possibly alert populations to the threat of locally triggered events, new technology is being developed at NOAA's Pacific Marine Environmental Laboratory. A new subsurface sensor has been designed to measure the pressure changes and timing associated with the passage of overlying waves, and to detect the specific characteristics indicative of a passing tsunami (*Figure 4.4*). Once a tsunami is detected, a surface buoy is triggered to send a signal via satellite to a shore-based warning center. When operational, these sensors will be able to confirm if a tsunami has actually formed and may provide a means of detecting locally derived events.

The new sensors are called bottom-pressure recorders, or BPRs. As a tsunami passes over the recorder (sitting on the sea floor), it measures the hydrostatic pressure changes associated with the passage of wave peaks and troughs. To detect a tsunami, it must measure pressure changes that are very small, but BPRs can record a 4-millimeter wave at a depth of 4,000 meters, equivalent to detecting a change of one part in a million—a truly impressive engineering achievement. Current work is focused on the difficult task of distinguishing tsunamis from wind waves and tides in the open ocean. Obviously, not every 1-meter-high wave in the open ocean is cause for concern, so the signals have to be carefully analyzed. Because a tsunami has such a long wavelength, the entire wave must go by before the data can be interpreted. This is not a problem when detecting a tsunami far offshore, but it is a great problem for one triggered and detected closer to the coast.

The wavelength of a tsunami is, in a sense, its most stealthy feature, for it also makes it invisible to satellite technology. Instruments aboard the TOPEX/POSEIDON satellite can detect wind waves on the surface of the ocean, but because

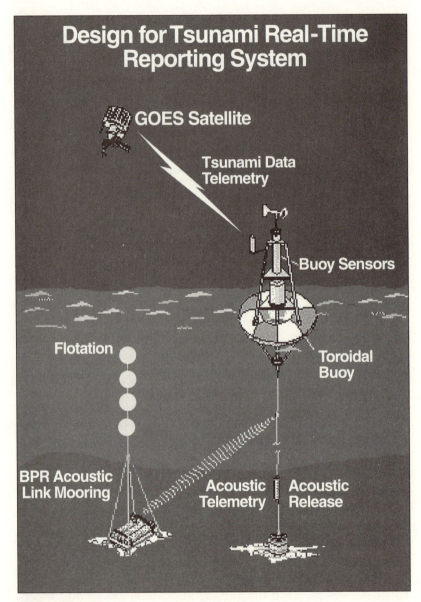

Figure 4.4
Diagram of open-ocean tsunami reporting system. *Courtesy PMEL/National Oceanic and Atmospheric Administration.*

the steepness of tsunamis is so small, they cannot "see" a tsunami. Wave steepness refers to the ratio of its height to its wavelength, and for a tsunami this is on the order of 1 to 10,000 or even smaller. Even if satellite detection was possible, a tsunami's racing speed adds to its stealthiness: an orbiting satellite would have to be in just the right place at just the right time to actually detect a tsunami.

In all respects, it seems that the new sea-floor pressure sensors offer the best hope for better and more reliable early warning systems. One had been deployed and was in operation for ninety days before being vandalized by fisherman. It is being replaced, and another four sensors are in the works. For effective warning, a much larger array of sensors is needed throughout the Pacific and off coastal areas at high risk. Eventually, the ultimate tsunami warning system will include a means to identify potential triggering events and detect or confirm tsunamis in the open ocean. Furthermore, each system will be linked to state-of-the-art tsunami models to predict where a tsunami will strike and the extent and force of coastal flooding expected, thus helping to determine where evacuations should take place. Some of these features are already in place, and others are under development.

When a tsunami is expected or has been detected, the Tsunami Centers issue a warning or watch, depending on how much time there is before the impact is expected. Warning messages are transmitted to the appropriate authorities and the public over the National Warning System teletype system and over commercial radio and television broadcasts. The NOAA Weather Radio system and the U.S. Coast Guard broadcast tsunami information directly to the public. Information on tsunamis is also now available via email and the Internet. Evacuation plans and orders are issued by local and regional authorities. The warning system is designed so that evacuees should only return to low-lying areas once the "all clear" is announced and the threat of tsunamis has passed.

For the tsunami warning system to be effective, local communities must be educated and prepared. Some tourism and financial corporations plan and educate employees about tsunamis. And in some areas, such as the Pacific Northwest, school children and the public are being instructed on where to go should an alert be issued. As described earlier, inundation mapping is providing crucial information for education and preparation for tsunami hazards. An excellent set of recommendations on how to respond in the event of a tsunami can be found in the USGS booklet entitled *Tsunami Survival*, which can be obtained from the USGS or state offices of emergency preparedness, such as the California Office of Emergency Services. At a minimum, people who live within the coastal zone should be aware of potential tsunami precursors or warning signs. An earthquake or a noticeable and rapid retreat of the sea should inspire those along the coast to move fast and far from shore and, if possible, up. And although false alarms are bound to occur, those who choose to reside in or visit tsunami-prone areas should never become complacent about tsunami warnings. History has illustrated all too well the tragic impact of tsunamis in regions ill prepared for this demonstration of the Earth's fury. And as one expert says, when it comes to tsunamis, we should learn to expect the unexpected.

Selected Sources, Readings, and Web Sites

The Underlying and Dynamic Earth

Baker, P., and McNutt, M. *The Future of Marine Geology and Geophysics: Report of a Workshop.* Ashland Hills, Oregon, December 5–7, 1996. National Science Foundation, 1998.

Broad, W.J. *The Universe Below: Discovering the Secrets of the Deep Sea.* New York: Simon and Schuster, 1997.

Decker, R., and Decker, B. *Volcanoes*, 3rd ed. New York: W.H. Freeman, 1998.

Dickinson, W.R. *A revolution in our time.* American Geological Institute. *Geotimes* 43(11):21–25,1998.

Hartmann, W.K., and Miller, R. *The History of the Earth.* New York: Workman, 1991.

Kennett, J.P. *Marine Geology.* Englewood Cliffs, N.J.: Prentice-Hall, 1982.

Kious, W.J., and Tilling, R.I. *This Dynamic Earth: The Story of Plate Tectonics.* U.S. Department of the Interior/U.S. Geological Survey, 1996.

Lamb, S., and Sington, D. *Earth Story: The Shaping of Our World.* Princeton, N.J.: Princeton University Press, 1998.

McPhee, J. *Assembling California.* New York: Farrar, Straus, and Giroux, 1993.

Monastersky, R. The globe inside our planet. *Science News* 154, 58–60, 1998.

Pinet, P.R. *Oceanography: An Introduction to the Planet Oceanus*. St. Paul, Minn.: West Publishing Company, 1992.

Press, F., and Siever, R. *Earth*, 3rd ed. San Francisco, W.H. Freeman, 1982.

Seibold, E., and Berger, W.H. *The Sea Floor: An Introduction to Marine Geology*, 3rd ed. Berlin, Heidelberg, Germany: Springer-Verlag, 1996.

Smith, W.H., and Sandwell, D.T. Global sea floor topography from satellite altimetry and ship depth soundings. *Science* 277(5334):1956–1962, 1997.

Woods Hole Oceanographic Institution, Massachusetts. Mid-ocean ridges. *Oceanus* 34(4), 1991/1992.

Web Sites

Here are just a few of the more general earth science resources on the Internet. Most of these sites have links to a wide variety of other relevant university and organizational links.

http://geoweb.tamu.edu

http://www.agu.org

http://www.enn.com

http://www.geotimes.org

http://www.globe.edu

http://www.ngdc.noaa.gov

http://www.sepm.org

http://www.usgs.gov

Earthquakes

Bartlett, K., and Wexler, D. News Notes. American Geological Institute. *Geotimes* 44(5):6–12, 1999.

Blakely, R.J.; Wells, R.E.; Yelin, T.S.; Stauffer, P.H.; and Hendley, J.W. *Airborne Hunt for Faults in the Portland-Vancouver Area*. U.S. Geological Survey Fact Sheet 097–95, 1995.

Bolt, B.A. *Earthquakes and Geological Discovery*. New York: W.H. Freeman, 1993.

Gere, J.M., and Haresh, C.S. *Terra Non Firma*. New York: W.H. Freeman, 1984.

Hauksson, E.; Jones, L.M.; Hutton, K.; and Eberhart-Phillips, D. *The 1992 Landers earthquake sequence: Seismological observations. Journal of Geophysical Research* 98(B11):19,835–19,858, 1993.

Hauksson, E.; Jones, L.M.; and Hutton, K. The 1994 Northridge earthquake sequence in California: Seismological and tectonic aspects. *Journal of Geophysical Research* 100(B7):12,335–12,355, 1995.

Hudnut, K.W.; Mori, J.J.; Prescott, W.H.; and Stauffer, P.H. *Southern Californians Cope with Earthquakes*. U.S. Geological Survey Fact Sheet 097–95, 1995

Hutton, L.K., and Jones, L.M. Local magnitudes and apparent variations in seismicity rates in southern California. *Bulletin of the Seismological Society of America* 83:313–329, 1993.

Kimball, V. *Earthquake Ready*. Malibu, Calif.: Roundtable, 1992.

National Geographic Society. *Restless Earth*. Washington, D.C.: National Geographic Society, 1997.

Shedlock, K.M., and Pakiser, L.C. *Earthquakes*. U.S. Department of Interior/U.S. Geological Survey, 1998.

Yanev, P. *Peace of Mind in Earthquake Country*. San Francisco: Chronicle Books, 1991.

Web Sites

These are just a few of the many wonderful Web sites on earthquakes, but most include a list and links to other relevant resources on the Internet.

http://geohazards.cr.usgs.gov/welcome.html
http://quake.wr.usgs.gov/
http://www.geophys.washington.edu/
    seismosurfing.html
http:/www.isc.ac.uk/
http://www.neic.cr.usgs.gov
http://www.seismo.com/iaspei/home.html
http://www.socal.wr.usgs.gov/seismolinks.html

Volcanoes

Bartlett, K., and Wexler, D. *News Notes*. American Geological Institute. *Geotimes* 44(5):6–12, 1999.

Decker, R., and Decker, B. *Volcanoes*, 3rd ed. New York: W.H. Freeman, 1998.

Dzurisin, D.; Stauffer, P.H.; and Hendley, J.W. *Living with the Volcanic Risk in the Cascades.* U.S. Geological Survey Fact Sheet 165–97, 1997.

Ewert, J.W.; Miller, C.D.; Hendley, W.; and Stauffer, P.H. *Mobile Response Team Saves Lives in Volcano Crises.* U.S. Geological Survey Fact Sheet 064–97, 1997.

Heliker, C.; Stauffer, P.H.; and Hendley, J.W. *Living on Active Volcanoes: The Island of Hawaii.* U.S. Geological Survey Fact Sheet 074–97, 1997.

Hill, D.P.; Bailey, R.A.; Sorey, M.L.; Hendley, J.W.; and Stauffer, P.H. *Living with a Restless Caldera—Long Valley, California.* U.S. Geological Survey, Fact Sheet 108–97, 1997.

Krafft, M. *Volcanoes: Fire from the Earth.* New York: Harry N. Abrams, 1993.

National Geographic Society. *Restless Earth.* Washington, D.C.: National Geographic Society, 1997.

Neal, C.; Casadevall, T.J.; Miller, T.P.; Hendley, J.W.; and Stauffer, P.H. *Volcanic Ash: Danger to Aircraft in the North Pacific.* U.S. Geological Survey Fact Sheet 030–97, 1997.

Newhall, C.; Hendley, J.W.; and Stauffer, P.H. *The Cataclysmic 1991 Eruption of Mount Pinatubo, Philippines.* U.S. Geological Survey Fact Sheet 113–197, 1997.

Press, F., and Siever, R. *Earth,* 3rd ed. San Francisco: W.H. Freeman, 1982.

Sorey, M.L.; Farrar, C.D.; Evans, W.C.; Hill, D.P.; Bailey, R.A.; Hendley, J.W.; and Stauffer, P.H. *Invisible CO2 Gas Killing Trees at Mammoth Mountain, California.* U.S. Geological Survey Fact Sheet 172-96, 1996.

Tilling, R.I. *Volcanoes.* U.S. Geological Survey, Denver, Colo., 1997

Web Sites

The list below includes just a few of the Web sites on volcanoes. Most include a list of links to other relevant resources on the Internet.

http://ca.water.usgs.gov/volcano/
http://hvo.wr.usgs.gov/
http://volcanoes.usgs.gov
http://www.avo.alaska.edu/
http://www.nasa.gov
http://www.noaa.gov
http://www.usgs.gov
http://www.usgs.gov/themes/volcano.html
http://www.volcano.si.edu/gvp

Tsunamis

Gonzalez, F.I. Tsunami! *Scientific American* 280(5):61, 1999.

Hokkaido Tsunami Survey Group. *Tsunami devastates Japanese coastal region.* American Geophysical Union. EOS Transactions 74(37):417, 432, 1993.

Monastersky, R.A. *Waves of death: Why the New Guinea tsunami carries bad news for North America. Science News* 154:221–223, 1998.

Pinet, P.R. *Oceanography: An Introduction to the Planet Oceanus.* St. Paul, Minn.: West Publishing Company, 1992.

Pond, S., and Pickard, G.L. *Introductory Dynamical Oceanography.* Oxford, England: Pergamon Press, 1983.

Satake, K.; Tsuji, Y.; and Ueda, K. *Time and size of a giant earthquake in Cascadia inferred from Japanese tsunami records of January 1700. Nature* 379:246–249, 1996.

Synolakis, C.; Imamura, F.; Tsuji, Y.; Matsutomi, S.; Tinti, S.; Cook, B.; Chandra, Y.P.; and Usman, M. *Damage conditions of East Java tsunami of 1994 analyzed. American Geophysical Union. EOS Transactions* 76(26):257–262, 1995.

Yamaguchi, D.K.; Atwater, B.F.; Bunker, D.E.; Benson, B.E.; and Reid, M.S. *Tree-ring dating the 1700 Cascadia earthquake. Nature* 389:922–923, 1997.

Yeh, H.; Imamura, F.; Synolakis, C.; Tsuji, Y.; Liu, P.; and Shi, S. *The Flores Island tsunamis. American Geophysical Union. EOS Transactions* 74(33):371–373, 1993.

Yeh, H.; Liu, P.; Briggs, M.; and Synolakis, C. *Propagation and amplification of tsunamis at coastal boundaries. Nature* 372:353–355, 1994.

Yeh, H.; Liu, P.L-F.; Briggs, M.; and Synolakis, C.E. *Tsunami catastrophe in Babi Island. Nature* 372:6503, 1994.

Web Sites

http://walrus.er.usgs.gov/docs/projects/
        cascad.html
http://wcatwc.gov
http://www.usc.edu/dept/tsunamis/
http://www.nws.noaa.gov/om/tsunami.html
http://www.pmel.noaa.gov
http://vulcan.wr.usgs.gov
http://www.geophys.washington.edu/tsunami

Endnote

Compared to our national investment in space science, our support of earth and marine sciences is embarrassingly meager. Given that we live on a planet whose surface is mostly covered with water, one would think that understanding the Earth and its oceans would be one of our highest national priorities. This is not to say that space exploration is unimportant; it has clearly provided us with a wealth of technology and significant information. But perhaps the time has come for us to give equal weight to the study of our own life-supporting planet.

Contributors

Dr. Kate Hutton is a seismologist at California Institute of Technology's Seismology Laboratory; Dr. Stanley Williams is a volcanologist at Arizona State University; and Dr. Costas Synolakis is a tsunami expert and professor of civil engineering at the University of Southern California.

Acknowledgments

My sincere gratitude goes to the U.S. Geological Survey and, particularly, Dr. Robert Tilling for the use of figures and information and Dr. Brian Atwater for his comments on tsunami research. Additional appreciation to Pamela Baker at the Ocean Drilling Program and Justine Gardner-Smith at Woods Hole Oceanographic Institution for their assistance in obtaining figures. Thanks is also extended to consulting experts Drs. Kate Hutton, Stan Williams, and Costas Synolakis. Dr. Synolakis specifically thanks Dr. Cliff Astill of the National Science Foundation, and Drs. Eddie Bernard and Frank Gonzalez for their advocacy, vision, and support of tsunami research. Special thanks to my editor at McGraw-Hill, Griffin Hansbury, for his guidance in seeing this book through. Last, but certainly not least, my gratitude goes to all my friends and colleagues who have supported my efforts to bring earth and marine science to the public.

Index

About the Author

Ellen J. Prager, Ph.D., formerly with the U.S. Geological Survey, is a respected scientist who is now devoting her time to bringing earth and marine science to the general public. In this endeavor she has been interviewed on the Today Show while living underwater, written dispatches for MSNBC, and worked with the National Geographic Society. She lives in Arlington, Virginia.